RETHINKING OUR WORLD

Dr Maja Göpel is a political economist and an important voice for a sustainable transformation of society, working at the intersection of the economy, politics, and society. From 2017 to 2020, she was secretary general of the German Advisory Council on Global Change, and in 2019 was appointed honorary professor at the Leuphana University of Lüneburg. She is a member of the Club of Rome, the World Future Council, the Balaton Group, the German government's Bioeconomy Council, and a co-founder of the Scientists for Future network.

RETHINKING OUR WORLD

an invitation to rescue our future

Maja Göpel

Translated by David Shaw

SCRIBE
Melbourne • London

Scribe Publications
18–20 Edward St, Brunswick, Victoria 3056, Australia
2 John St, Clerkenwell, London, WC1N 2ES, United Kingdom
3754 Pleasant Ave, Suite 100, Minneapolis, Minnesota 55409, USA

First published in Germany by Ullstein Buchverlag as
Unsere Welt neu denken in 2020

Published by Scribe 2023

Typeset in Adobe Caslon Pro by the publishers

Printed and bound in the UK by CPI Group (UK) Ltd, Croydon
CR0 4YY.

Scribe is committed to the sustainable use of natural resources and the use of paper products made responsibly from those resources.

978 1 957363 36 3 (US edition)
978 1 925322 61 3 (Australian edition)
978 1 911617 42 6 (UK edition)
978 1 761385 11 7 (ebook)

Catalogue records for this book are available from the National Library of Australia and the British Library.

scribepublications.com
scribepublications.com.au
scribepublications.co.uk

For Juna and Josephina —
my two amazing daughters

Contents

1

An Invitation

'In the middle of the twentieth century, we saw our planet from space for the first time. Historians may eventually find that this vision had a greater impact on thought than did the Copernican revolution of the sixteenth century, which upset the human self-image by revealing that the Earth is not the centre of the universe.'

From the United Nations' Brundtland Report[1]

London, October 2019. During the morning rush hour, two men clambered onto the roof of a tube train, preventing it from leaving the station. The commuters wanting to board the train to continue their journey to work were confronted with closed carriage doors. The interruption soon paralysed the entire tube line, and the station platform grew ever more crowded and noisy. As it slowly dawned on the increasingly angry crowd that they were now bound to be late for work, the men on the roof of the train unfurled a banner bearing the slogan *Business as usual = death*.

For these commuters, 'business as usual' meant going to work in an office, perhaps, or a factory. Sitting down at a computer, or in a meeting, or at a machine, to manufacture something or to order the manufacture of something. To increase sales and profits, to contribute to growth, to secure their own job and their own economic future. To pay their rent, service their debts and buy themselves and their children something nice. In short: to continue their lives just as they, and all of us, have become accustomed to.

What can be so wrong, or even deadly, about that?

The two men protesting on the roof of a tube train that autumn day in London belonged to a group of activists calling themselves Extinction Rebellion.* The 'extinction' they were 'rebelling' against is not just the demise that we have come to accept of a rapidly increasing number of animal species. The men were not just concerned with whales, bees, or polar bears. Without a trace of irony, they meant the extinction of our own species, the human race. They meant us.

Compared to Greta Thunberg—the girl whose school strike triggered one of the most widespread protest movements in human history—the members

---

* Here my aim is simply to give an example of non-violent civil disobedience as a form of protest. I expressly dissociate myself from individual statements made by the leaders of this movement in England.

of Extinction Rebellion are the civilly disobedient ones among climate protectors and environmentalists. Like others, they demand that politicians and policy-makers finally take sustainable action to stop and reverse global warming, for which they also offer specific proposals. But rather than just joining marches, they deliberately set out to disturb public life, often wearing colourful costumes, and always with the basic rule of remaining friendly at all times. On that autumn day in London, hundreds of activists blocked streets, chained themselves to bridges, and glued themselves to an airport floor. Using as little violence as possible to make as great a public impact as they could, their aim was to disrupt in a tangible way what they see as the true cause of climate change and the rampant destruction of life on Earth: our normal, everyday life.

For the people who were prevented from boarding their train that morning, the situation was so intolerable that they began pelting the two activists with sandwiches and drinks. Eventually, when that failed to remove them, one of the commuters climbed up and dragged the men off the roof of the train and onto the platform, where the activists were beaten by the angry mob until the police were able to intervene and arrest them.

That confrontation was not about a nourishing piece of bread, a drink of clean water or a protective roof over somebody's head, or even the last litre of petrol. All that was at stake was a couple of minutes' delay on people's way to work. One side wanted to save the world; the other wanted to get to the office. One side wanted to break with old habits; the other wanted to cling onto them. Although we must recognise that both sides were basically concerned for their future and that of their children, their concerns appeared to be mutually exclusive. It appears one side has to lose for the other to win. It can only be either 'us' or 'them'.

Is that what the future looks like, in times of climate change?

Is this how our lives, and our future struggles, will be?

In the world we now live in, systems that have for decades supplied us reliably every day with ever more energy, food, medicines, and security are coming under pressure everywhere and almost at the same time. They shaped an epoch in which, generally speaking, increasing amounts of everything were constantly available; prosperity, even for the poor; progress in every branch of science and technology; and peace, even between countries with fundamentally differing political systems. When there is a constantly increasing supply of everything, questions of distribution become less important. The fact that people are astonished at the idea that this epoch might come to an end, that they resist even the thought of this happening, and that they are at a loss to imagine what might come next, shows how accustomed we have become to this situation, how normal we consider it to be. What was thought of as a privilege by our parents' generation has become routine for most people today.

At the same time, we know that 'business as usual' is not going to work.

It's not just about climate change, plastic in our oceans, burning rainforests, or factory farming. It's also about exploding housing costs in our cities, out-of-control financial markets, the ever-growing gap between rich and poor, the increasing incidence of burnout syndrome, and the incalculable and multilayered consequences of gene technology, and digitisation. For some time now, our perception of the world has been coloured by a feeling that times are changing. Our present seems fragile, while our future appears to be heading unstoppably for those scenes we associate with apocalyptic sci-fi films. The utopias so touted by the Modern Age have become dystopias. Our confidence in the future has turned into worry and fear. The solutions that worked so well on a small scale, and that promised such comfort, turn out to be threats when scaled up to the global level. We have a sense that we are about to face immense upheavals.

It has become increasingly difficult to explain our immediate future in terms of our recent past; everything that could be relied upon, all the tried-and-tested answers, fall apart. Every answer to one problem seems to aggravate another. And so disagreements over which problems should be addressed first are bound to increase. But what if we found ways to tackle multiple problems at once? Ways that might involve questioning old certainties, but that allow us proactively to shape a future we want to live in, rather than simply reacting to each negative situation as it arises?

So my invitation to you is to come along with me to explore those ways. The future is not something that just comes out of the blue. It's not something that just happens. In many ways, the future is the result of the decisions we make in the present.

That's the reason I'd like to invite you to take a closer look at the world that is my, your, and everybody's home, to rediscover what is possible in our world. People have done this many times through human history, typically in times of crisis. Many technological breakthroughs were born out of the need to find alternative solutions. This is happening today in the case of renewable-energy sources. Many fundamental societal changes were born out of a conviction that things could be run differently. And, lo and behold: women can vote, and even head national governments.

The scale of the upheavals we currently see means they do not just affect certain sections of societies, but societies in their entirety. Academics describe them as major transformations in economic, political, social, and cultural processes, which also means they alter the way we see the world. The Neolithic Revolution, or the much later Industrial Revolution, are often cited as examples of such upheavals in the past. The former refers to the adoption of a sedentary lifestyle by small groups of nomadic people, later leading to the development of a feudal, agrarian society. In the latter, the use of fossil fuels in particular opened the way for a complete reorganisation of economic activity and of society itself, leading to the growth of a bourgeois middle class and the rise of the nation-state.

Our modern world differs fundamentally from that of 250 years ago, when the Industrial Revolution began. And yet we still usually try to find solutions with the worldview of that time. We have forgotten how to assess whether our ways of thinking are still fit for purpose in our times. But learning to question those thought patterns will reveal the tools that will enable us to escape this crisis and shape the future in the twenty-first century.

This is book is not a 'climate book'. It is not about how many degrees the average temperature of the global climate will rise by in the coming years, and the consequences that will have for life on our planet. It doesn't tell of melting ice sheets, constantly rising sea levels, or regions that can no longer support human habitation because they are now under water, have been turned into deserts, or are regularly visited by devastating storms. This book doesn't describe the largest mass extinction since the demise of the dinosaurs, or the acidification of the oceans, or drought, famine, epidemics, forced mass migration, or any of the other countless scenarios that scientists all over the world have been warning us about for decades and that are now becoming real far more quickly than even the researchers who document them thought possible.

I am not a climate scientist. I am a social scientist, and my main academic interest is political economy. I study the ways people organise their economic and social activities, their relationship with nature, and their interactions with other human beings. I examine their use of natural resources, energy, materials, and labour. I look at the rules people adopt when organising work, trade, and financial exchanges. I scrutinise the technologies people develop and how they put them to use.

But, most of all, I seek to understand why the specific solutions that arise do so, and why some ideas become established in society while others do not. What ideas, values, and interests lie behind those ideas? Where do these ideas come from? How do they develop into the powerful theories that dominate not only our modern economies, but also our thoughts, actions, and lives in general—and sometimes even our feelings? And why are the ideas that have been perpetuated in these theories for the past 250 years no longer necessarily helpful as we strive to turn the current ecological and social crises into opportunities for the future?

It may feel as if our economic system developed naturally, just as plants and animals once evolved without our interference. But human-made systems work in a different way. We assess the situation we are in and then establish rules to change that situation. Such changes may affect cultures, or markets, or they may be as simple as the demarcation of a national boundary, but they are usually more complex in nature. In our daily lives, we barely notice this creative part of our reality and are scarcely aware of its history, because what were once ideas and innovations have long become truisms, laws, institutions, and customs. But it remains the case that they are self-imposed rules that now make up the world we have created for ourselves today.

If we want to understand how humanity could have brought the planet — the only one we have — to the brink of collapse in the space of just two generations, we become aware once again of those ideas, structures, and rules.

But what does 'become aware of' mean here?

It means recognising what we are doing and asking why we are doing it. Social scientists call this approach 'reflexivity'. It offers an opportunity to learn. If we don't question our actions and our motives, we will not be in a position to act differently. If we are not open to alternative solutions, our response to new problems as they arise will often be simply to repeat our previous actions.

Asking radical questions and experimenting with different solutions is a way to regain freedom and creativity. It offers an opportunity to create in good time new, original responses to challenges, rather than always meeting them with long-established, reduplicated answers. That is why I love being a scientist. And it's also the reason I wrote this book. It is not a compendium of nuanced details, facts, figures, and comparisons between individual models and forecasts. It is an attempt to present the main strokes of the transition we are currently experiencing in as approachable a form as possible, and to offer some ideas and perspectives to mediate between the seemingly intractable positions of the preservationists and obstructionists, so that we can gain some sense of direction in our search for a sustainable future for us all.

I grew up in a village near Bielefeld, Germany, in an old farmhouse renovated by my parents and some of their friends who also had children. The house was big enough for each family to have its own space, but we spent most of our time together. Those children of my parents' friends are still like siblings to me today. We all attended the same school—a new, progressive school that did not award grades, replacing them with what they called 'learning process reports'. The adults took turns looking after us after school so that the others could work. We kids spent much of our time playing in the old construction-site trailer that was parked in the garden, which we, of course, had painted with all the colours of the rainbow. Unsurprisingly, the rest of the village saw us as hippies, although all the grown-ups had middle-class jobs. My parents are medics, and campaigned for better disease prevention and treatment of trauma. Both were, and still are, members of the organisation IPPNW—International Physicians for the Prevention of Nuclear War.

Mine was a typically alternative childhood for West Germany in the 1980s. But the broad range of backgrounds among the children at our progressive school meant I was constantly reminded of how privileged we were, growing up in our eco-social farmhouse. I wasn't particularly keen on the veggie burgers we often had at home, or at least I would have liked a glass of Coke to go with them, but that was not on our menu. I didn't particularly miss meat, but I did bemoan the lack of milk, nuts, and mushrooms. This was the time just after Chernobyl. I still remember the big sack of powdered milk we kept in the pantry, and being told we couldn't go out and play in the fields on the first few days after the accident. No one knew how high the radiation levels were. It was strange, especially because the threat was invisible. A few years later, the first Gulf War came along, and we protested for peace with other school students by blocking Jahnplatz,

Bielefeld's central square. At some point during that period, I began asking myself: if everyone I know wants to create a world of love and peace, bring an end to poverty, and ensure the environment is nice and safe, why don't we just do it?

What is stopping us, as a society?

I suppose the quest to find answers to that paradox is what drives me, even to this day. I have studied in Germany, Spain, Switzerland, and Canada, backpacked round South America and the US, done voluntary work for the German Federation for the Environment and Nature Conservation, the BUND (the German branch of Friends of the Earth), leading to visits to Hong Kong and Mexico and to my participation in world trade conferences in cooperation with the international network 'Our World Is Not for Sale'. Working with pioneers of sustainability from all over the world, I have helped develop political guidelines for the 'World Future Council' (WFC) foundation, to improve the protection of the rights and interests of future generations, and promoted them at the United Nations in New York and the European Union in Brussels.

When I became a mother, I chose to settle professionally at the Wuppertal Institute, a centre for environmental, climate, and energy research. Working there, I have been able to connect much of my practical experience with transformational research approaches and develop them further on a theoretical level. I have always had one foot in the world of academia, but I never wanted to acquire knowledge only to share it with a small circle of fellow experts and decision-makers. I've always also felt drawn to broader society, mostly to places where people are passionate about striving for goals beyond their own prosperity and success, and are determined to do all they can to achieve that aim. I have learned an unbelievable amount from such people, and have tried to incorporate that passion into my academic work.

Between 2017 and 2020 I was the secretary-general of the German Government's Advisory Council on Global Change (WBGU). That's a body of independent experts that regularly gathers information on the state of academic knowledge concerning the most important trends in environmental and development science, to make that knowledge accessible to political decision-makers. I spend most of my time communicating those results to the public and making them easily understandable to as many people as possible. It is especially important to me, in the 'post-truth' era we now supposedly live in, to stick steadfastly to my faith as a humanist in the power of science and knowledge. And to my belief in the opportunities offered by communication and understanding when attempting to get to the roots of any misunderstanding, when people can approach each other in a space that goes beyond the rigid, inflexible roles they normally play.

This was my motivation for setting up the group 'Scientists for Future' (S4F), initially with a small group of scientists, and writing an open letter expressing my complete support for young people's street protests, and justifying them with a series of facts. We never dreamed that 26,800 scientists and academics from Germany, Austria, and Switzerland would sign that letter within three weeks. Or that the nationwide press conference held to explain our position would go viral on social media.

We see it as our responsibility to provide innovative alternatives as we move through these changing times.

I also see people's willingness to inform themselves and to question old certainties as an amazing opportunity. That doesn't mean there is a quick answer to the paradox I recognised when I was young, but it does create the most important condition for change—a space where the idea of change becomes possible.

The global environmental and social crises we see today are not random. They reveal the way we treat ourselves and the planet we live on. We will not overcome these crises without an understanding of the rules upon which we have based our economic system. Only when we become aware of those rules will we be able to change them—and win back our freedom.

2
A New Reality

'Science is part of the reality of living; it is the what, the how, and the why of everything in our experience. It is impossible to understand man without understanding his environment.'

Rachel Carson, scientist[1]

On the morning of 21 December 1968, the US astronauts Frank Borman, William Anders, and James Lovell took off from the Kennedy Space Center in Florida, heading for the Moon. Their mission was to orbit Earth's natural satellite and to photograph the lunar surface, gathering information for a later Moon landing. Apollo 8 was also scheduled to fly behind the Moon, to the side that constantly faces away from the Earth, which no human had ever laid eyes on, and so the three men were expected to return to Earth with an entirely new image of the Moon.

They were already on their fourth orbit, and the nose of the spacecraft was about to re-emerge from the Moon's shadow, when the commander turned the ship, whose nose had constantly pointed towards the unfamiliar lunar surface, and, suddenly, the Earth rose into view through a side window.

'Oh my God!' exclaimed Anders, who was the first to notice the sight. 'Look at that picture over there! There's the Earth coming up. Wow, that's pretty!'

The cockpit radio recording is available online. In the audio, Anders can clearly be heard frantically asking his crewmates to find a colour film to replace the black-and-white one in his camera, followed by their repeated questions to confirm he really had snapped the image.[2]

'You sure we got it now?'

'Just take another one, Bill!'

The picture Anders took shows a bright-blue ball, marbled with swirling white clouds, with parts of the beige and green of the continents occasionally visible beneath. This is our home—a small, almost fragile-looking planet surrounded by the bottomless blackness of space—the only planet in the solar system that is home to life.

Frank Borman, William Anders, and James Lovell set out to gain a new view of the Moon. What they came back with was a new vision of the Earth. The photo they took was later published by NASA under the poetic title *Earthrise*, and is still considered not only one of the most important photographs in human history, but also the most powerful environmental image of all time. The reason is simple: it shows our entire environment in one single picture. This is the only planet we have.

The image didn't show anything that hadn't essentially already been known for at least 500 years. The fact that the Earth is not flat was common knowledge by the time of the first circumnavigation of the planet, at the latest. The idea that the Earth does not form the centre of the universe—and that humanity is therefore not the centre of everything—was a long-accepted concept. But the finite and unique nature of this globe had never before been seen so startlingly clearly. Our everyday view of the world tends not to take in the larger connections and interrelationships.

The image people have of something does not necessarily say anything about the thing itself. First and foremost, it tells us something about people's attitude to it. There's a difference, and it's a huge one. In fact, it's so great that it is the source of almost all the problems we face today.

When the Apollo 8 mission set off for the Moon in late 1968, there were around 3.6 billion people on Earth. By the end of 2019, our fragile celestial body was home to more than 7.7 billion people. In the space of just fifty years, the Earth's population had more than doubled. Glib mention of that figure is often made, and such statistics constantly come up in the context of global population growth, but what are we supposed to make of these figures? An increase from 3.6 to 7.7 billion in fifty years: is that fast or slow?

A comparison might help.

If you imagine the history of humankind as a feature film—from just over 300,000 years ago, when the species *Homo sapiens* first appeared in Africa, up to the present—most of the movie would be over before humans even adopt a sedentary lifestyle and develop agriculture and animal husbandry. By the time the world's population reaches 1968 levels, the film is almost finished; then, in the last few seconds before the credits roll, the figure suddenly doubles.

In other words: at mind-blowing speed.

But that's not even the point.

It's not just that there are twice as many people alive on Earth now as there were fifty years ago; most of those people also take up considerably more space than their ancestors, especially in countries that have been particularly successful from an economic-development point of view. Anyone who doubts this just needs to cast their mind back to how their own families lived fifty years ago, or quiz their parents about it.

Where did they fly to for their holidays? Abroad? And how many times a year? Did they even fly? Or drive? Did they own a car? Two cars? How big was their home? Did each child have their own room, their own TV? How many clothes did they have in their wardrobe? How many electronic devices in their home? And how much of the household technology that we now take for granted didn't even exist? How often did your family buy themselves new clothes? How often did they replace the furniture? And were those things manufactured in, and shipped from, countries far away?

In short: what was normal for people fifty years ago, and what is normal for us today? How many factories, power plants, roads, airports, and industrial farms were necessary to provide their normality, and how many are necessary for ours?

One economic indicator that measures the impact a person's lifestyle has on our planet is their ecological footprint. It includes not only the amount of agricultural and pasture land it takes to produce an individual's food, the roads they use, and the land they live and work on, but also, for example, the amount of forest necessary to fix the carbon dioxide released into the atmosphere in the production of the energy that person uses. A person's ecological footprint converts the natural resources they consume into a number of hectares, and compares that with the amount of space available in nature to counterbalance that consumption, so that what has been harvested can grow back, or so that the natural environment can recover. Any gardener knows you can't reap more than you grow. The idea of the ecological footprint takes this rule, expands it by the factors that make our world more complex than a gardener's allotment, and applies it to the entire planet and to humanity as a whole.

When the Apollo 8 crew took off for the Moon, humankind's ecological footprint was still within the bounds of what the Earth can sustainably deliver. It has been consistently above that level since the mid-seventies. Overconsumption of the Earth's resources has become a permanent condition. Each year, the date when humanity's demand for ecological resources and services exceeds what the Earth can regenerate in that year comes earlier and earlier. In 2019, that date fell on 29 June. Every day after that date, we are borrowing more from nature than we can pay back, leaving us with fewer environmental resources for the following year than ever before. This date has been labelled 'Earth Overshoot Day'; for Germany alone, it fell even earlier than for the world as a whole—on 3 May.

As one of the world's largest exporters, Germany imports natural and industrial resources from other countries on a grand scale. If the prevailing lifestyle in Germany were to become the global norm, we would need more than two entire Earths. But, as the famous photo snapped by the Apollo 8 astronauts shows, we only have the one. Still, we regularly see protests against personal or legislative curtailment of our consumption whenever this uncomfortable truth is brought up.

So if we are to shape a successful future for ourselves, it must be based on the current state of the world, and not on the way things used to be. For thousands of years, people's experience of the Earth was that of a planet with unlimited resources. If the forest was cleared in one place, they just moved on to the next part of the woods. If all the game had been hunted, the lake fished empty, or the mine's deposits exhausted, they simply relocated, or moved their reliance to a different resource that was still in good supply at the same location. Our planet felt gigantic. It was always possible to avoid shortages one way or another, or to move on to virgin land. Of course, that was not always a peaceful process—far from it. In particular, as the nation-states of Europe were just beginning to establish themselves and expand around the globe, they dispossessed those living in newly 'discovered', less densely populated regions and continents, and often totally decimated them.

The increasingly prosperous industrialised nations gained access to immeasurable new resources, and developed new technologies or discovered previously undreamed-of building blocks of existence, such as atoms and genes. What we call modern progress is, when it comes down to it, nothing more than expansion and extraction. As long as this model worked well, as long as there were few people and a lot of planet, there was no reason to change it. Struggles for social justice and universal human rights may have continually changed the methods of such progress, but the principles it was based on were by and large never questioned. However, the relation between humans and the natural world has changed fundamentally. There is now less and less planet available to more and more people. Our economic activity no longer takes place in an 'empty' world, but in a 'full' one, to paraphrase the economist Herman Daly.

This is nothing short of a new reality.

But what does that mean?

It means that the coordinates of human coexistence and successful economic activity have fundamentally shifted. Expansion and extraction come to a natural end when nature and its ecosystems are robbed of the ability to regenerate reliably. Scientists call these 'tipping points' or 'planetary boundaries'. If we want to live in the real world—and one whose reality is currently undergoing radical change—we must recognise these boundaries. Otherwise, we are simply living in a make-believe world.

Since this new, twenty-first-century reality that has arisen between humankind and the Earth is a global one, it also changes the life of every human on the planet. Pretending that this isn't the case is simply living in an illusory reality of global proportions—but that's precisely what usually happens in the debate about the climate crisis and sustainable development. We may talk of planetary boundaries, but most proposed solutions avoid recognising their significance. Pay attention, and you will realise that the talk is almost always of more growth and more prosperity—but rarely of where that growth and prosperity will come from, and at what cost.

It has now been fifty years since the publication of one of the earliest warnings that humanity must adjust to this new reality if it is not to sleepwalk into a catastrophe of global proportions. It came from a group of scientists led by Dennis and Donella Meadows who, working at the Massachusetts Institute of Technology in Boston, pioneered the use of computer simulations to make predictions about the future of humankind. The model they developed was named 'World3' and would run on any home computer now, but required an entire mainframe to run back then. The scientists fed their model with data on five long-term factors. These concerned the pace of growth up to the present in the world's human population; global food production; industrial output; human exploitation of non-renewable resources such as metals and fossil fuels; and pollution. And, most importantly, they investigated the way these five growth trends interact with each other.

The researchers' aim was to extrapolate the data on past developments to make forecasts about the future.

The scenario they called their model's 'standard run' was based on the assumption that there would be no major changes in human behaviour or values, and that we would continue along the lines of 'business as usual'.

When their results were published in 1972, the impact couldn't have been greater, even if they had predicted that a giant asteroid was about to strike the Earth.

Indeed, in a sense, that *is* what they predicted.

Their calculations showed that under standard-run conditions, human civilisation was heading inexorably for collapse — and would reach that point within the next hundred years. If the global population and industrial output continued to grow, the forecast continued, non-renewable resources would soon start to run out, while the harm caused by pollution would reach irreparable levels. The costs of these developments would exceed what the system could withstand. The system would begin to wobble, industrial output would start to fall, and the population would start to shrink. At some point, the curves of all five factors would tip steeply into the negative, one after the other; hence the term 'tipping point'.

Even more shocking was the fact that the scientists couldn't stop their computer model from predicting a system collapse even if they controlled some of the factors. If, for example, they switched the non-renewable resource parameter to 'infinite', the population grew so much that it outstripped the amount of agricultural land needed to feed everyone. If they limited population growth and doubled the amount of food produced, the resulting increase in pollution began to push up the mortality rate. Whatever they tried led sooner or later to the same result.

The only scenarios that did not end in collapse were those in which the growth of all five factors was limited. That appeared to be the only way to avert system breakdown. This is why the scientists called their report *The Limits to Growth*.

Essentially, the scientists had not recognised anything that wouldn't be obvious to anyone with eyes and a sense of logic. But since many local environmental problems in rich countries were solved by improving technology and by relocating environmentally harmful processes to other countries, it was possible to record and process global interconnections only with the advent of modern computer technology. What had once been pure thought-experiments now became visible, quantifiable graphs. That's what propelled the study to having such a huge impact.

The study is still famous today: its findings have been updated and re-evaluated many times, but they have never been fundamentally refuted. In essence, all five factors have developed more or less as the scientists predicted almost fifty years ago. This should come as no surprise. After all, humanity has not deviated from the standard-run scenario, even after seeing the threat of the collapse of its formula for progress printed in black and white. It has simply carried on as before. Relative gains in efficiency, and improvements in individual products and technologies, have not changed the overall picture. There has still been no conclusive, absolute decoupling of the growth of a national economy from the environmental degradation that results from that growth.

There have been repeated attempts since the seventies to get to grips with this problem—not only its individual manifestations, but in its entirety. Efforts have been made to describe the problem, to raise awareness of it, and even to solve it. Fresh studies have been carried out, councils and commissions have been established, summits have been held, and reports have been written and officially approved. However, a glance at just one manifestation of the problem—the battle against climate change—is all it takes to see just how much, or how little, progress humankind has made on this front.

The fact that carbon dioxide emitted into the atmosphere causes the Earth to warm up, and that humans are accelerating that process by burning fossil fuels such as coal, oil, and natural gas, has been proven and accepted by science since the late nineteen-thirties. In 1957, American scientists warned their government that human beings were now unwittingly 'carrying out a large-scale geophysical experiment'.[3] By the late seventies, scientist already knew almost everything we know today about climate change. Since 1992 and the advent of the United Nations Framework Convention on Climate Change, there has been an international agreement—now ratified by almost every country—committed to slowing down global warming. Since 1997 and the Kyoto Protocol, there have even been legally binding targets for the reduction of greenhouse-gas emissions, which were made more ambitious in Paris in 2015 with a further agreement aiming to limit global warming to well below two degrees Celsius.

And if the idea of anthropogenic global warming wasn't common knowledge before, it became a widely known fact with the release of Davis Guggenheim's documentary film *An Inconvenient Truth*, about the former US vice president Al Gore's campaign to prevent climate change—a film that garnered two Oscars and helped Gore win the Nobel Peace Prize for his commitment.

And that was in 2007.

Did you know that half of the carbon dioxide released by human activity has been emitted in the past thirty years?[4] That means by us, by our generation. The damage our generation has caused, in full knowledge of the consequences, has now reached the level of that caused by humanity in the entire time before we knew what we were doing.

How could this happen?

My theory is that we have simply refused to acknowledge the new reality. For almost fifty years now, we have lived in a pseudo-reality that we created for ourselves by following monetary rather than physical and biological indicators.

For a long time, we lived in a situation where there was a lot of planet shared by few humans. But now there are more and more humans and less and less planet to support them. If humankind is not to bring about its own collapse, we must learn to manage in a crowded world, on one single planet with limited resources. This is the new reality.

3

Nature and Life

'If a society cannot deal with resource depletion, then the truly interesting questions revolve around the society, not the resource. What structural, political, ideological, or economic factors in a society prevented an appropriate response?'

Joseph Tainter, anthropologist[1]

In March 2018, the US patent office received an application to patent a new technology for pollinating plants artificially. In the document, covering several pages, the authors of patent application number US2018/0065749 describe their invention of a very small 'aerial vehicle' similar to a mini-drone, which can leave its docking station automatically and fly unpiloted over farmland. Using a tiny brush, it can collect pollen from one plant and, by means of an equally tiny fan, apply that pollen to a different plant. A sensor detects whether the pollination has been successful, and sends a signal to a network hub so the same flower does not get pollinated again by another 'vehicle'.

There are at least two things that may surprise readers of this patent application. First, it is easy to see that the inventors have engineered a replica of something that has existed in nature for millions of years: the bee.

But in the world described by the inventors in their patent application, something seems to have changed. They write that the populations of insects that plants normally rely on for pollination have been in steady decline for years. The authors also point out that attempts to pollinate crops using blanket-spraying of pollen from planes have proven ineffective.

The second surprising thing about this patent application is that it wasn't filed on behalf of the inventors themselves, but for the company they were employed by: the American retail giant Walmart.

Why should a retail company be interested in robot bees?

Well, Walmart is not just any chain store; it's the world's biggest retailer and one of the financially strongest companies overall. Walmart achieved this with a simple business model: to be cheaper than its competitors at any cost. 'Always low prices, always' was its advertising slogan for many years. This strategy means that Walmart makes less money on the sale of each individual product than do many of its competitors, so it has to sell a tremendous number of each item to turn a profit. This is an example of what social scientists call 'economies of scale': bulk selling brings in the profits.

That has made Walmart not only the world's largest company by revenue, but also the world's largest private employer, with more than two million employees working in over 11,000 locations. It should come as little surprise that the Walton family, who founded the Walmart business, have been ranked the richest family in the US for many years.

But what does this have to do with artificial bees?

In order to understand how and why our economic system developed into what it is today, we must also understand our attitude to nature. The natural world is the basis for all our economic activity, providing both the necessary energy and materials. We humans don't create them; we simply convert them. When people believed the natural world was created by one or more gods, the laws of nature were seen as just as unfathomable as the will of God or the gods. Some cultures framed Nature or the Earth itself as the creative goddess. Western culture eventually settled on the idea of a God who created the Earth and gave custodianship of it to Man.

From the sixteenth century, scientists and

thinkers such as Galileo Galilei, René Descartes, and Isaac Newton cast new light on those concepts and reinterpreted the idea that we are destined to be 'the masters of the Earth', and a completely new perspective on the role of humans emerged. Such scientists proved that the natural world *does* follow predictable rules, from which it followed that, if science could recognise and describe those rules, and humans could systematically put them to use for our own benefit, our fate would finally be in our own hands. This resulted in the Enlightenment and in a whole new way of the *Homo sapiens* species seeing itself.

Like a child taking a toy apart, humans dismantled nature bit by bit, and started playing with the parts. They learned the role that each component plays. They altered the components, exchanged them, or reassembled them in new ways, in the conviction that this would make the world better suited to their purpose. Humans' view of nature, which they had been a part of previously, changed, as they began to see it simply as 'the world around them', from which they had completely separated themselves and that now simply formed their surroundings. What had been a living whole whose parts were all interconnected came to be seen as a machine that humans could alter and rebuild at will to suit their needs. What had been a kind of dynamically self-stabilising network of relationships was reduced in humans' perception to a set of individual elements, and often to the one particular aspect of the (now invisible) whole that was of interest to them.

And what was of interest to them was whether it could be used to make a profit.

Or, they asked, if not, can we get rid of it?

Anyone who sees the world that way is, of course, oblivious to the incredible diversity of nature, the dynamic way it changes, and the interconnections between its individual parts. Such a view blinds us to the fact that no one thing, not even the tiniest of snowflakes, is ever identical to another in nature. And to the fact that every natural phenomenon is the result of other phenomena, and that the way an element is embedded in nature affects its quality and its development. The resulting view of the world goes something like this:

The forest is nothing but timber.

Soil is a support matrix for plants.

Insects are pests.

And a chicken is a thing that provides eggs and meat.

Every chicken that has ever been kept by humans throughout their history has been a descendant of the red junglefowl, a free-ranging wild species originally native to South and South-East Asia, which was domesticated by humans and spread around the entire globe. It is now the most abundant species of bird in the world. However, the breeds we keep today have almost nothing in common with the wild form, and our modern chickens differ enormously even from those kept by our ancestors around 100 years ago. Before then, it was usual to rear chickens that both laid eggs and could provide meat, although some breeds were better suited as sources of one or the other. When farmers tried to improve one trait through selective breeding, the other would be negatively affected. More eggs meant less meat, and vice versa.

Since the end of the Second World War, humans have divided these animals according to those traits, and created breeds that are suitable only for one use or the other. Modern broiler chickens are ready for slaughter after just one month of growth; modern laying hens produce up to 300 eggs in their first year of life—and a second year is not part of the plan for them. But a far worse fate awaits the male chicks of the egg-laying breeds. In this system, they are doubly useless, as they naturally don't produce eggs, nor do they put on meat quickly. Raising them makes no economic sense, so they are sent to the shredder as soon as they hatch.

Do you find that perverse?

This is the system that produces 12 billion eggs a year in Germany alone, with 650 million chickens slaughtered—and 45 million chicks shredded.[2] And then the same again the next year. (In 2022, Germany became the first country in the world to ban this practice by law.)

Chickens that used to be able to provide everything, living on farms that would utilise everything they could provide, have been replaced in the course of the development of modern civilisation by highly optimised chickens in highly specialised poultry factories. The various aspects of chicken farming have been completely separated from one another. Today, there are companies that specialise in breeding, while others only concentrate on rearing; yet others only engage in fattening, and there are those that exclusively produce eggs. Although centuries of selective breeding by humans led to great diversity, the chickens used in this system come from very few breeds. This genetic downsizing has created animals that are far more susceptible to disease. The lack of diversity is also reflected in the ecosystem of poultry and egg producers: organised into quasi-monopolistic structures, with the market dominated by a small number of companies, they can be, and have been, bankrupted by a single wave of bird flu.

It's a similar picture when it comes to cash-crop farming, in which highly saleable produce—such as bananas, coffee, soybeans, or wheat—is grown to generate financial profit, rather than for local consumption. Cash crops are grown for export, and often take precedence over a country's own needs.

Cash-crop farming uses highly efficient varieties to produce the greatest possible yield in the shortest possible time. Unfortunately, such varieties tend not to be particularly resistant to changes in climate, but the practice has already led to the disappearance of most alternative varieties.

The crucial difference between the kind of systems that modern humans create and those found in nature is that the latter show high levels of diversity, and function in a cyclical way. In naturally occurring systems, unlike in human-created ones, no one removes elements without returning them in a form in which they can be re-utilised by that system. The waste products of some parts of the cycle are the food of others. When humans intervene in such naturally developed systems, the cycle becomes a conveyor belt running in only one direction. At the front end of the belt, resources are removed from the system and consumed, and at the other end, the belt spits out waste that is food for no one. That waste must be burned, buried, dumped, or simply left to accumulate in rivers or seas.

Natural systems are designed to last; human systems are created for the moment. Natural systems thrive on diversity, are self-regulating, and are able to absorb shocks. Those characteristics are precisely what makes them resilient and efficient as a whole. They are geared towards energy efficiency, and so let nothing go to waste. Modern, human-made systems concentrate on the economic efficiency of individual processes. Think of my conveyor-belt analogy: lower costs at the front end mean a net positive result at the back. As a result, human systems reduce diversity to create a homogenous overall structure, thereby rendering that structure fragile and prone to faults. Rather than adopting the patterns of successful evolution found in living systems, modern humans attempt to transform everything they get their hands on into a maximally productive machine without considering the environment that 'machine' is in.

And it's not only nature that we treat this way.

The next time you take a stroll through a town centre, take note of how many small, independent shops you see, compared to the number of international chain stores selling the same things as in any other city, or country, or continent. Clothing is a good example: garment manufacturing creates 92 million tonnes of waste every year. That 'rubbish' often includes perfectly functional garments, but they are usually burned, simply because that's the cheapest way to get rid of them.[3] This means that another grab for the Earth's resources will be necessary for the next mass-produced fashion collection, rather than recycling what already exists.

Or have you perhaps long-since stopped even going to your local town centre, since you can do practically all your shopping on Amazon anyway—that amazing mega-company that offers everything more cheaply and more conveniently? However, people are becoming increasingly aware that Amazon uses its retail activities to monitor and analyse our entire society, and that it sells such information for a profit. As is the fact that this gigantic internet platform systematically targets brands and manufacturers that refuse to sell their products via Amazon.

What is only now coming to light is the fact that the movements of its warehouse workers are directed by wristband scanners that sound an alarm when the workers take longer than the standard time to complete a task. Self-employed delivery drivers barely see another human, even during the recruitment process. It's all just emails, videos, and navigation devices. In addition, Amazon pays barely any tax, as it only declares profits in a few places that impose little corporate tax in order to make themselves attractive business locations. But the company is perfectly happy to use taxpayer-financed infrastructure in all countries, or to exploit social welfare systems that support those in precarious employment. Even the cycle of 'a portion of my profits goes to maintain our public services' breaks down in this model.

This now-global model of progress as a mechanical engine of extraction and maximisation has subjected not only the natural world, but also cultures and ways of life to a rapidly advancing process of homogenisation and economisation. All around the globe.

There are almost 2.5 billion active users a month on Facebook.

Starbucks, Zara, Primark, McDonald's, Burger King, and Coca-Cola manufacture and sell their goods everywhere.

We all watch the same movies, listen to the same music, recognise the same stars, and eat burgers, pasta, and pizza. Worldwide.

But what does all this have to do with robot bees?

In 1983, the United Nations set up a commission to consider how our economic activities could be reconciled with the natural limits of our planet. The commission was chaired by the former Norwegian prime minister Gro Harlem Brundtland. Its report was therefore known as the Brundtland Report, and was published four years later. For the first time, it set out guidelines for the direction that human economic activity needed to take for it to be sustainable. The idea was to create a simple point of orientation for redressing the ecological balance. Even back then, it was already clear that things were starting to get out of kilter.

The commission's definition of sustainability was simple, and has since formed the basis of all international environmental agreements: 'Sustainable development is development that meets the needs of the present without compromising the ability of future generations to meet their own needs.'[4]

This is followed by two important sub-points: that overriding priority should be given to the needs of the poor; and that care should be taken to ensure that social and technological developments do not disrupt nature's regenerative cycles. This was a call for a paradigm change.

Nineteen eighty-seven was also the year that the US economist Robert Solow won the Nobel Prize for his model of economic growth, which not only considered the role of new inventions as a driver of national economies, but also included the idea of the substitutability of natural capital. This sounds far more complicated than the Brundtland Report's rules for sustainable economic activity, but in fact is equally simple.

Intead, it approaches the search for solutions from the opposite direction. The substitutability of natural capital is based on the idea that it is possible to remove individual elements from a natural system and replace them with artificial equivalents. So, according to Solow, it would not be a disaster, or even a mistake, if humans destroyed nature, as they could simply replace it with technology, and everything would be fine. Green becomes grey. This constituted a reinterpretation of the second condition in the Brundtland Report: social and technological processes no longer needed to be integrated into nature in such a way that they did not disrupt its regenerative cycles. Now, all that humans needed to do was to find suitable replacements.

Or, in the sober words of Robert Solow himself, 'If it is very easy to substitute other factors for natural resources, then there is in principle no problem. The world can, in effect, get along without natural resources, so exhaustion is just an event, not a catastrophe.'[5]

This first time I read this, I could hardly believe it.

And they gave him a Nobel Prize for that?

Important institutions such as the World Bank adopted this view, and heaped praise and money on countries that paid for education, housing, and other things by exploiting their natural capital. This was called the 'genuine savings' approach. According to this metric, the disappearance of the rainforests is not a problem as long as people make a lot of money out of the products and services the forests yield, since the only economic parameter being recognised is money or price. But a monetary indicator cannot show whether human-made substitutes will be able to integrate into the network of life. The issue of whether it is okay to just destroy all life, as long as we can build machines to replace it, has remained strangely undiscussed within the supposedly value-neutral discipline of economics.

As you may have noticed, I found Solow's view to be presumptuous, and its basic assumptions to be pretty devoid of any scientific knowledge. By contrast, the Brundtland Report expresses a far more real-world view. But, leaving that aside, the approaches of Solow and Brundtland only epitomise two ways of seeing the world—as has so often been the case throughout the history of mankind. They offer two different options for the future: to carry on as before, but in a more extreme way; or to make fundamental changes. If you change the way people view the world, the world itself will change. There was a choice between these two options in the 1980s, and that choice is still the same today.

Can you guess which view has prevailed since the showdown in 1987?

Which brings us back to the robot bees.

The pollination of plants by insects can be seen as a service provided to humans by nature. Converted into monetary terms, Germany's Federal Agency for Nature Conservation estimates the value of pollination services worldwide to be more than 150 billion euros a year.[6] That's more than the annual profits reported by Facebook, Apple, Alphabet (Google's parent company), and Microsoft combined. Other services provided for humans by natural ecosystems include the purification and circulation of water, air, and nutrients; protection against storms and flooding; and the recreational value of natural spaces. This makes estimating the total value of all the services provided by ecosystems very complex—after all, it is an attempt to show the amount of added economic value provided by nature in our lives, as compared to human-made forms of value creation. Or conversely, the question can be framed as how much it would cost if we had to create that value ourselves—quite apart from the question of whether we would even be capable of doing that.

A 2014 meta-analysis by a group of researchers led by Robert Constanza came up with a sum so huge as to make minor deviations in either direction negligible: by 2011, nature was providing humans with eco-services to the value of $125–$145 trillion a year. That figure is significantly larger than the global gross domestic product (GDP)—that is, the total value of all goods produced and services provided by people in one year. Global GDP stood at $84 trillion in 2018, whereas in 2007 it was only around $55 trillion. The study also places the yearly cost of the destruction of ecosystem services at $4.5 to $20.1 trillion in the period up to 2007.[7] If we weigh up the growth in GDP against the cost of the destruction of the ecosystem, the balance is negative.

Although ecosystem services are of such immense value for the reliable and healthy provision of resources and a high quality of life, humankind receives them practically free of charge from nature. We don't have to invent or develop them, or pay for people and machines to do so. That's why they don't appear as items on our balance sheets, and, since anything that does not have to be paid for is not recognised as having economic value, nature has simply never been given sufficient consideration up to now. We pay for the individual pieces, the units of resources that we take from the Earth—such as a cubic metre of timber or a gram of iron, for example. But we have no functioning pricing system for, let alone an understanding of, the regenerative and distributive purification of air or water, the dispersal of pollen or seeds, the capture and sequestration of carbon dioxide, or the safeguarding of food chains and biodiversity. Are you beginning to see how strange it is to think of protecting the natural world and engaging in successful economic activity as being opposed to each other?

A third of the world's crop production relies on pollination by insects. But as long as companies such as Walmart are concerned only with selling foodstuffs as cheaply as possible, they will of course remain oblivious to the damage caused by precisely the kind of industrialised agriculture required to produce that food so cheaply.

Fortunately, such companies are increasingly coming to recognise this.

For several years now, Walmart has been making efforts to become a sustainable company. It has modernised its huge fleet of heavy vehicles, reduced the electricity consumption of its refrigeration systems, and reduced its packaging sizes, thereby cutting a lot of carbon dioxide emissions that would otherwise have contributed to accelerating climate change. When Walmart then started installing solar panels on the roofs of its ridiculously large supermarkets, it became the largest producer of solar power in the US. It even added organically produced goods to its range, which instantly turned it into the biggest purchaser of organic milk and organic cotton in the world.

That sounds like a great success story, right?

Surely, when such a huge company suddenly starts operating sustainably, the whole system will end up being geared towards sustainability? Well, you might think so. But the economic concepts of growth, productivity, and competition that I intend to describe and question in this book have prevented precisely this from happening. There has been no such change in either the company's operations or the markets for milk and cotton.

Walmart has not turned into the world's largest chain of organic supermarkets. Instead, it is paying people to develop robot bees.

Whether those mini-drones really will fulfil the function of bees or not, it is a reckless experiment, to say the least. Amazon, for example, still needs human workers because robot hands are not yet dextrous enough to do all its warehouse jobs. And microelectronic devices are quite fragile; they are not nearly as robust as a self-repairing biological bee. Furthermore, all these human-made technological substitutes require a source of energy, which must also be provided by humans. This stands in opposition to the fact that we are even now trying to reduce energy consumption to curb climate change. Bees make their own energy from their food. They live off the pollen that they collect from plants, and from the honey they produce themselves. Plants obtain their energy through photosynthesis, which functions without any human intervention at all and without causing any damage to other ecosystem services.

I'm sorry, Mr Solow, but even if we reduce the ethical issues and value judgements to nothing more than the survival of 'team humanity', the idea of creating a future economic system whose entire functionality depends on human-made mechanical processes and energy sources is, when it comes to resilience, sheer madness.

Why not simply preserve the natural world we have been given, with its diverse energy sources and ability to regenerate? We already know which farming and planting methods are decimating the populations of real bees. So, which innovation program is likely to be more life-sustaining? Mini-drones, or a transformation of farming methods, supply chains, and land-use concepts?

Our increasingly common attitude to the natural world reveals the great arrogance at the basis of human economic activity. Subjecting natural systems to our needs reduces nature's diversity, makes it more vulnerable, and requires ever-increasing efforts to stabilise it. Human systems are not sustainable, and will inevitably collapse if we don't learn to change them.

4

Humans and Behaviour

'When an idea becomes successful, it easily becomes even more successful: it gets entrenched in social and political systems, which assists in its further spread. It then prevails even beyond the times and places where it is advantageous to its followers.'

John Robert McNeill, historian[1]

The ultimatum game is a scientific experiment to study human behaviour. It was devised in 1982 by the German economist Werner Güth and his colleagues. First, they gave one of two test subjects a certain amount of money. That person was then asked to share the money with the other test subject and to decide on the proportions of the split. Under the rules of the game, the proposer could make only one offer; no follow-up negotiations were allowed. If the responder accepted the offer, both parties got to keep the money they ended up with. If the responder rejected the offer, both left the game empty-handed. That meant the first subject had to be very careful to make an offer that they thought the responder would accept.

The researchers found that there appeared to be a minimum cut that the proposer had to be willing to give up for the responder to accept the offer. The share was around 30 per cent of the original sum. So, for example, if the proposer was given 1,000 euros to start with, they had to offer to hand over at least 300 euros; otherwise, the responder would reject it.

Not surprised? Well, it did surprise the economists.

If we are to rethink our world, we must start from the fundamental concepts on which the world we are familiar with today is built. They include humanity's view, not only of the natural world, but also of itself. You might think that humans would not be quite as far off the mark with this as we are with our view of nature. When it comes to the human race, surely humans must be the experts? Unfortunately, the opposite often turns out to be the case.

The view of human beings that lies at the core of most economic theories is one of a cool and calculating egoist who is out to gain a personal advantage from every situation. When people are faced with a choice as consumers, they will always choose whatever results in the greatest personal benefit for themselves; as producers, they will always opt for whatever promises the greatest profit. Feelings—both one's own and those of others—play no role at all in such decisions. Cold reason is the only factor, and even that rationality is reduced to a cost-benefit calculation. Economists call this perfectly rational, self-interested actor *Homo economicus* ('economic man'), and that is also the name of the economic concept that has long been used to explain humans' economic behaviour. Of course, *Homo economicus* was intended as a roughly drawn theoretical symbol, but it has been taken as the basis of many an economic model.

That's why the results of the ultimatum game experiment were so surprising to those who study economics. If the responders in the game were acting as *Homo economicus* should, they would have accepted any amount they were offered. Irrespective of how small the amount was, *Homo economicus* would never pass up such a benefit. However, the fact that the test subjects preferred to go empty-handed when they felt the proposer was not sharing fairly seemed illogical, according to the economists' model. It contradicted their view of human beings and the widespread theoretical models based upon it.

* * *

But why is it that sustainable societies are still so difficult to achieve? As a young adult, I assumed, naïve as it may sound, that it was simply down to a lack of knowledge. I thought that once people realised they could behave in a different way and learned how to do that, they would be able to create sustainable societies. So I decided to enrol at university to study media. But, as you have seen, it's important to ask what knowledge actually is, and what kind of knowledge is helpful.

A behaviour that seems 'logical' to the vast majority of us is considered by the vast majority of those who teach and study economics at the top universities to be a deviation from the—well, I can only call it rather sad—norm of human existence. This fact was astonishing to me. And I was even more astonished when I had to attend some economics lectures to attain an extra EU qualification to enhance my degree, and learned more about the theories with which economists view the world. Suddenly, I realised that this depressing view had become the mode for a phantom world. Real people were as absent from the economic theories as was the real natural world. Essentially, everything came down to businesses striving for ever-greater profits and households aspiring to buy ever more goods and services, to fuel ever more growth for national economies. The only value recognised in this view was money.

I remember, in one lecture, the professor telling us that workers will always travel to wherever wages were higher, even if that means moving to another country. When I raised my hand and asked him just how bad local poverty has to get, and how big the difference between earning potential at home compared to abroad has to be, for people to be willing to leave their families, and why the economic models don't include the costs of such an effort to the workers in question, silence descended over the lecture theatre.

The professor turned to his assistant, and the other students all stared at me. Finally, he said, 'See, those are the words of someone with a warm heart!'

I never got an answer to my question. Since then, I have been preoccupied with the question of why economics is so proud of its cold-heartedness, and why that should be a desirable trait. But it did make me feel as if I'd taken an important step towards answering the question of why we are incapable of creating sustainable societies. I decided to write a doctoral thesis on the history of ideas in economics, to explore critically the way that this phantom world has come about, and to examine the role its ideas have played in political and social development.

The way economists assess human behaviour, and measure whether their economic behaviour is rational or not, goes back to work of three men. All of whom were born more than 200 years ago, and all in Great Britain. As such, it is hardly surprising that the economic system they based their view of human beings on—namely that of industrialisation—also originated in that country. After all, theory and practice do not usually develop in isolation, but tend to reflect each other.

The first of the three men is Adam Smith. His book *The Wealth of Nations*, which was published in 1776, is still a much-cited work today. According to Smith, individuals employ their labour to produce whatever they are best at producing. The resulting variety of products can be traded on the free market, where prices are dictated by supply and demand. Thus, according to market logic, the self-interest of each individual results in benefits for all—as if guided, as Smith puts it, by an 'invisible hand'. It's an almost-magical image, which, however, came to be seen as far more important by later interpreters of his work than it was for Smith himself.

The second of the three men is David Ricardo. He raised the concept of the division of labour and exchange to the level of entire nation-states. He developed a theory of international trade that asserted it is advantageous for every state to trade with other states, even if the goods that one state produces are also produced in the trading-partner state, possibly even at a lower cost. Ricardo used the example of Portugal and England to illustrate his model. At that time, each country produced both cloth and wine, but Portugal was able to produce both goods more cheaply than England. Ricardo showed that it still made sense for the two countries to trade with each other, because Portugal required fewer workers to produce wine than England did to produce cloth. So, if Portugal specialised in wine production and England concentrated on cloth-making, both countries would be able to produce more overall than if each country produced both. This so-called comparative advantage is still the basis, or rather the justification, for international trade today.

The third of the three men whose work forms the basis of our economic models was not an economist at all, but a naturalist. Charles Darwin recognised that new species arise through the processes of random genetic change and natural selection, and that natural selection is correlated with the ability to adapt to change. The then-nascent discipline of economic science applied this perspective to its own subject matter—with its supporters led principally by the philosopher and sociologist Herbert Spencer. Suddenly, economic theorists and policy-makers were no longer concerned with finding a sensible way to organise the division of labour so as to produce an increasing number of goods to satisfy people's needs. At the interpersonal level, economic activity was now seen as an every-man-for-himself fight—a fight that only the strongest would survive.

If you follow these three assumptions, engaging in economic activity is nothing more than an attempt by egotists to survive in a world full of other egotists by producing ever-increasing amounts of goods and services, which results in them accumulating wealth, as well as more prosperity for all miraculously coming about.

How does that sound to you?

Like a story with a catch?

Then you probably still have in mind *Homo economicus* and what became of him in the ultimatum game. Because maybe the story isn't true — even though we constantly hear it told in its many variations in the business and economics reports that appear in the media we consume.

In the mid-seventies, the American economist Richard Easterlin published an essay with the title 'Does Economic Growth Improve the Human Lot?', in which he compared economic data gathered from nineteen countries over a period of twenty-five years. He correlated that data with the results of surveys conducted in those countries to determine how satisfied citizens were with their quality of life. He discovered that, at a certain point, the average happiness at a national level stops increasing as average per-capita income grows. While there was initially a reliable correlation between the per-capita figures for gross domestic product and gross domestic happiness, there appeared to be a level at which that correlation breaks down, when more prosperity did not automatically lead to a better quality of life.

This contradiction was named after the man who discovered it, and is still known as the Easterlin Paradox today, although it does not really seem like a paradox to anyone who isn't an economist to hear that accumulating more and more does not automatically lead to more happiness. When we have enough to eat and drink and have a roof over our heads, matters such as good health, sound relationships, a fulfilling occupation, and affirmation from others naturally gain in importance in our subjective evaluation of our quality of life. And yet the cleverest minds in economic science still continue to break their heads over questioning the concept of *Homo economicus* and the market and social developments modelled on his behavioural repertoire. The fact that those models (and computational models) have always been founded on *one* representative actor, whose decisions form the basis for forecasting the dynamics of the economy, means it is not easy to make the models more closely representative of the real world.

Social scientists call this 'methodological individualism', and most economists so far continue to adhere to this principle. It focusses on human decisions about the use of the means—which are always defined, as a matter of principle, as scarce—to achieve a goal—which is also always defined, as a matter of principle, as ... I think you can see what's coming ... increased consumption. Only now are economists slowly beginning to develop what they call agent-based models, which take into account differences in character, but which are therefore far more complex and really require a lot of computing power.

And economic theories are, of course, always gross simplifications. They have to be. A theory is first and foremost just a certain, specific way of interpreting the world. Economic theories are based on single aspects of reality and a deliberate decision to emphasise those aspects over others, while ignoring other aspects completely. That is not a flaw; on the contrary, it is necessary if theories are to fulfil their intended purpose—that of creating clarity in a world that seems anything but clear, before they are supplanted by another theory that may fulfil that purpose even better.

Of course, there is something to the phenomena that led Adam Smith to conjure up the image of the 'invisible hand'. But all too often, economists forget the fact that he made his observations at a time when small craftsmen's businesses and the cottage industries of England only traded among themselves. Globalisation, dominated by huge, multinational business concerns, did not exist. And the fact that Adam Smith's first great work was entitled *The Theory of Moral Sentiments*, in which he describes the ability to feel 'sympathy' (which we would probably be likely to call 'empathy' today) as an essential human trait, is ignored or suppressed just as often as the fact that Smith clearly advocated for regulatory laws, and by no means assumed that the market would regulate itself.

Or take David Ricardo. He could not have known that, one day, there would be financial markets in which capital could move freely around the world with no regard for the conditions of production in any given country. Modern international trade is no longer about having a small number of trading partners and specialising in a few products; its scale is now global. A country that participates in free trade is automatically in competition with every other participating country. The relative cost differences for individual products have been replaced by absolute cost differences between entire national economies in providing the basic conditions for production. There is increasing pressure on countries to reduce their production costs so that the sales prices of the products on the global market can also be reduced, at the expense of social and environmental concerns. Comparative advantage culminates in a constant battle to make everything, everywhere ever-cheaper. This is what economists call competition.

And Charles Darwin? Evolution is a trial-and-error selection process, but it always results in more diversity rather than less. Of course, there are always stronger and weaker individuals, but the decisive factor is their ability to adapt themselves and the ecological niche they occupy. However, if we assume that the prevailing conditions are better for a few organisms than for others, the requirement to 'be generally fitter' differs according to those conditions. Also, competition in nature is always limited to a local area, rather than operating worldwide and creating monopolies for the victors. This makes sense, as, when conditions change, it is best to have as many options open as possible. That's why ecological niches and the organisms that occupy them, or the solutions they offer, are so valuable when it comes to the continuity of the whole and the emergence of the new.

All three of these intellectual pioneers have had their central ideas taken out of context by their successors and elevated to the supposed universal law of 'the' economy.

Why is it important to realise this?

Because economic science is not just something dreamed up by a few professors living in a world of their own and producing studies that nobody reads. On the contrary, their scientific hypotheses form the basis of company balance sheets and business models, for economic policy-making, for public institutions, and, whether we like it or not, for our own behaviour. Economic science informs the values on which we judge whether something is efficient or not. It defines our concept of progress.

For a long time now, hasn't the charge that something is 'inefficient' been one of the most devastating verdicts anyone could make about it?

And isn't the incredible increase in prosperity we've enjoyed since the Second World War confirmation of fact that we simply must follow the findings of economic science?

People have always based their lives on theories, on insights gained by thinking about so-called reality. But if a theory provides only a distorted description of reality when it is put to the test, it's not simply a problematic theory. If we follow the theory too slavishly, the eventual result will be the production of a new reality out of that theory. Or a pseudo-reality.

This is why reflexivity is important in science: the aspiration of science to constantly update its theories. And if that reflection reveals that an entire operating system no longer works, then it must be updated.

Or would you be okay with bringing up your children according to the pedagogical rules in place over 200 years ago?

Homo economicus recognises no qualitative difference between different resources; no difference between genders; no cooperation; no empathy; no feelings of responsibility, either at the level of the individual or society; and, strictly speaking, he doesn't even recognise such a thing as society at all. No one is born a *Homo economicus*; but, being a social species, humans can be trained in that direction when they are brought up in a system that constantly rewards them for behaving like *Homo economicus*. The theory becomes the practice. We all like to find good stories that make our behaviour seem plausible, or at least legitimate, to others. Selfishness, ruthlessness, and cold-heartedness are then no longer the characteristic traits of humanity, but simply the result of an education that suppresses qualities such as altruism, the ability to share, and warm-heartedness.

Jamie Gamble, a former corporate lawyer to some of the biggest companies in the world, describes this with regard to business corporations, and his conclusion can be summarised as follows: the exclusive focus on corporate stock value means that corporate managements and boards of directors are obliged by law to act like sociopaths. Companies' relationships with their employees, with their customers, and with the communities in which they produce and sell, as well as their effects on the environment and on future generations, have no real place in this economic philosophy.[2]

But even beyond the corporate world, this economic way of thinking has infiltrated areas of our lives that originally had nothing to do with economics. The provision of care—for the sick, for old people and for children—has been yoked into this logic, as have the education sector, our choice of partners, and even our own bodies. For example, paediatric wards in Munich hospitals are being closed because treating children simply takes too long. Hospitals receive a fixed payment per child treated, irrespective of how long that treatment takes. So the less time spent listening, explaining, or comforting, the greater the profits.

When we take a vacation, our holiday has to be both relaxing and exciting at the same time; we don't have much time, after all. When we have children, our offspring have to be successful in life, otherwise all the time and energy we invest in them will have been for nothing—and, of course, in this value system, 'successful' is more likely to mean earning a huge income like that of an investment banker, rather than choosing a career that adds social value, such as becoming a midwife.

When we turn on the television, we see casting shows in which contestants offer themselves up like goods to the harsh judgement of the market (the viewers). And if we take up yoga or meditation to avoid burn-out from all that stress and pressure to perform, our aim is not to come to a realisation about all this and find a way to escape the rat race. No, we take up such practices so we can get back to full performance capacity more quickly, and become even more focussed, more productive, more attractive. This is called self-optimisation, and it will hopefully become more convenient as it is increasingly automated with digital devices and implants. After all, we are all human capital and must be sure to maximise our value on the market.

Social media are not the only—but perhaps the most visible—places where we see that the concept of selling and competition has infiltrated areas of our lives in which the law of supply and demand used to be less important than intrinsic values. We hear of people who are only able to gain a sense of self, of their existence in the world, by constantly googling their own name and checking how many followers, likes, or friend requests they have.

How can we escape this situation?

We can see what can happen when just one basic premise of a theory is changed by looking, for example, at the way work is understood in Buddhism. In the economic models of the West, with its concept of progress, work is seen as a cost borne by employers, which they always want to minimise. Workers are seen as having to forfeit their freedom and free time to engage in labour, for which they must be compensated by the payment of their wages. As such, an ideal world for both sides would be one in which employers no longer have to pay workers anything, while workers receive pay for doing nothing.

In Buddhism, on the other hand, work is seen as something that helps develop a person's abilities. It connects people, and stops them becoming lost in self-centredness. In addition, labour is a means of producing the goods and services that are necessary and desirable for a dignified existence. In this approach, an ideal world would not be one in which production is increased at as little a cost as possible, but would be what we call an inclusive 'activity society', which ensures the welfare of everyone. Instead of relying on automation, it focusses on human activity, supplemented by technology when people want help with their work. There is a difference between a tool that increases a person's strength or skill, and a machine that takes work away from them. Organising work so as to produce as many goods as possible, as quickly as possible, at all costs, would be an offence according to the Buddhist worldview, as it would be prioritising production volumes over people, and products and profits over human relationships and experience.

Have you noticed something?

In order to rethink the world, it is sometimes enough to think differently about just one thing.

Buddhist economics was described by the British-based German economist E.F. Schumacher after he spent time in the mid-fifties working as an economic advisor in the country then known as Burma. His book *Small Is Beautiful: a study of economics as if people mattered* was already one of the most influential works on sustainable economies before the term even existed. Published in the early nineteen-seventies, it quickly became a bestseller. The future scenario it describes feels like an answer to questions we are still asking ourselves today.

Schumacher was never awarded the Nobel Prize.

Even now, the leading economic journals almost never run articles that question their own worldview. So I was all the more impressed when the Organisation for Economic Cooperation and Development (OECD) held a conference entitled *Averting Systemic Collapse*, in September 2019. A small OECD think tank called New Approaches to Economic Challenges (NAEC) published an additional report. It gathered together a long list of empirical results showing the inadequacies of the *Homo economicus* model, and demonstrated that both the concept of the substitutability of capital and the idea that economic growth is synonymous with increased inclusivity, justice, and quality of life are unhelpful when it comes to our interactions with the natural environment.

The group had scarcely published its findings when the national representatives from the US spoke up to remind the NAEC's program director that such ideological aberrations were not compatible with the founding principle of the organisation. After all, it is the paying member states who determine the mandate of the OECD. And the majority of company CEOs? According to Jamie Gamble, their reaction to the suggestion that corporations should bear legal responsibility for their impact on their employees, customers, regions, the environment, and future generations has also been less than euphoric.

Schumacher once wrote, 'I certainly never feel discouraged. I can't myself raise the winds that might blow us or this ship into a better world. But I can at least put up the sail so that when the wind comes, I can catch it.'[3]

And the OECD did raise at least a little wind—perhaps even gusting against that veto from the US. Whatever the case, the OECD then changed its motto from 'Better Policies for Growth' to 'Better Policies for Better Lives'.

Most people in economic science still see humans as selfish creatures who look out only for their own advantage and, by doing so, miraculously create prosperity for all. This view of human beings is wrong, and urgently needs updating. A system that rewards selfishness produces selfish people. We need to reconsider the values that support humans in their cooperative vigour.

5
Growth and Development

'The world is facing three existential crises: a climate crisis, an inequality crisis, and a crisis in democracy. Yet the accepted ways by which we measure economic performance give absolutely no hint that we might be facing a problem.'

Joseph Stiglitz, Nobel Prize-winning economist[1]

Karsten Schwanke is one of the meteorologists who present the weather every evening before the prime-time news on the first channel of Germany's public-service broadcaster, ARD. Apart from forecasting the weather, the show also includes short items on meteorological phenomena of interest to general viewers. In just three or four minutes, Schwanke is able to explain to audiences why rainbows are bent, for example, or why clouds don't fall out of the sky. Although such questions are unlikely to have occurred to viewers before, they usually find they are fascinated by the answers. Some time ago, Schwanke began dealing with questions about climate change in this segment of the show. He explained, for instance, why Antarctica's ice is melting, although the temperature there never rises above freezing, or how a drought in Germany was connected to forest fires in California and flooding in Italy. Suddenly, the apocalypse was cropping up in an early-evening TV segment that usually dealt with something as innocuous as the weather. It was almost as disconcerting as two men climbing onto the roof of your underground train during London's morning rush hour, just when you needed to get to the office.

Karsten Schwanke's segments explaining climate change are a huge hit on social media. The videos continue to get thousands of shares and millions of views, even months after being broadcast. ARD even receives requests from viewers to turn the format into a full-length early-prime-time program before the nightly news.

As a sustainability scientist, I have also joined those calls, because that would help the topic gain more traction, and embed it as relevant information in our daily consciousness. As a political economist, on the other hand, I'm particularly charmed by the idea that the daily climate report would then be broadcast right after the business and stock market news—going from graphs about stock-price increases straight to graphs about increases in carbon dioxide emissions.

That would make the climate costs of our economic system visible in a direct, graphic, and compact way to prime-time TV viewers.

The Mauna Loa Observatory in Hawaii has been measuring the amount of carbon dioxide in the Earth's atmosphere since 1958. The location was deliberately chosen for its distance from any centres of civilisation, and its position on the leeward side of a volcano at an altitude of more than 3,000 metres above sea level and almost 4,000 kilometres from mainland America—all with the aim of avoiding any distortion in the data collected. That information, which has been gathered continuously for more than sixty years, is among the world's most valuable data sets.

The curve representing those measurements plainly rises almost continuously. There are only three exceptional places where it flattens slightly—once in the mid-seventies, once in the early nineties, and once after 2008.

What is significant about those times?

The mid-seventies was the time of the oil crisis, when Arab states reduced their oil production by just 5 per cent, causing the price of oil to almost double in a short space of time. The beginning of the nineteen-nineties saw the collapse of the Soviet Union, and the global financial crisis of 2007–08 caused the growth rate in several countries' GDP to slow down. Politically all very different, those three events had the same economic effect: a drop in production, transportation, and consumption, and therefore also a reduction in carbon dioxide emissions.

In other words: when the economy shrinks, climate change slows down. When the economy grows, climate change speeds up.

Or to put it even more simply: economic growth as we know it today is synonymous with climate change. And more economic growth equals more climate change.

That is the fatal logic on which our civilisation is based.

You don't believe it?

Then just compare the graph from Mauna Loa with a graph tracing global economic output over the past sixty years. You will not only see both curves constantly rising; you will also see that the total cuts in carbon dioxide emissions achieved so far have not been sufficient to change the overall trend. The two curves are almost exactly congruent, as the physicist Henrik Nordborg showed in his essay 'A Spectre is Haunting the World—the Spectre of Facts'.[2]

This is one unpleasant observation that we must face

up to. The other is the realisation that all our attempts to break this link have so far failed.

Neither international climate accords such as the Kyoto Protocol and the Paris Agreement, nor the development of renewable energies have been able to stop the continued increase in carbon dioxide levels in the atmosphere.

And what about the graphs for raw materials extraction, deforestation, biodiversity loss, or plastic waste, for example? It's the same pattern everywhere: the trends are all identical—the lines on the graphs curving upwards like hockey sticks.

It's depressing to see it summarised like this, but it's not really much of a surprise. As long as humanity continues to cling to the idea that economic production must always grow, any progress we make in improving conditions for ourselves and our environment in one area will be more than cancelled out in another.

Might this not be due primarily to the rapid growth in the world's population in the same period? Yes, partly. But in Germany, for example, the population has not grown significantly for decades, and for a period it was even shrinking. The country has been a leader in climate protection, but that's mainly because the collapse of the former East Germany's industry meant a sudden and massive reduction in carbon dioxide emissions. Yes, there have been a lot of technological improvements, and much progress has been made in recycling, which has led to significant improvements in the total amount of energy and resources used in proportion to economic output. Fridges, cars, and radiators no longer guzzle as much energy as they used to. Nonetheless, the overall demand for electric power has risen by more than 10 per cent since 1990, while energy consumption has fallen by only around 3 per cent.[3]

That's precisely why the forecast made by the authors of *The Limits to Growth* in 1972 still applies today: growth in economic output is limited, because there is a natural limit to the amount we can take from and add to the planet. Despite this, we continue to quantify economic performance—by which we mean growth—without regard to those looming physical limits.

The measure we have come to use for this, gross domestic product (GDP), includes only the value added through the production or sale of goods and services in a country each year. When the concept of GDP was invented in England over 300 years ago, it still retained the distinction between land, livestock, and state treasuries. However, the concept was not used in a targeted political way until the Second World War, when the US, in particular, sought a better understanding of how quickly its economy could grow to produce the necessary level of armaments. Since then, the annual GDP figure has been used as *the* economic indicator of growth, and therefore prosperity. Here, as always, a concept has become a number, and that number has become the basis for decisions, policies, and the orientation of society as a whole. The amount of value lost and environmental damage behind the number remains hidden.

Examples?

When an oil slick from a tanker accident pollutes a section of the coast, the result is an increase in the GDP, because it creates business for companies that scrape the oil off the beach, as they are rendering a service for which they are paid. The damage to the ecosystem caused by the oil spill is not reflected in the GDP figure, because—as we have seen—nature never figures in economic balance sheets as long as it has no positive or negative economic (or monetary) effect. By contrast, fathers or mothers staying away from the office for a period after the birth of their child will cause the GDP to go down. The wellbeing of the baby and its parents as they start their life together is not counted in the GDP figure.[4] Perhaps the most striking definition of this economic indicator stems from John F. Kennedy's brother, Robert Kennedy, when he said in 1968 that it 'measures everything except that which makes life worthwhile'.

Despite this, it is assumed in most economics textbooks that the bottom line is positive on the whole. Of course, that's connected with the concept of *Homo economicus*, who, as we know, is not only egotistical but also insatiable. So individual benefits come from more consumption or less work.

Just to reiterate: when the world was relatively empty, with few people, little material wealth, and lots of nature, it was quite reasonable to assume that producing a lot more would result in lots of benefits. The economic system we have built upon this idea is oriented towards production, in order to create growth and to then invest that growth in such a way that innovations lead to yet more production. Increased production means increased benefits for consumers. In the old reality, as I'd like to call it, in which the majority of people had to manage on very little wealth, or none at all, it was easy to see the logic behind this equation, which forms the core of our idea of economic progress. The equation still holds true for countries that lack access to good-quality food, safe accommodation, adequate clothing and healthcare, or a reliable energy supply.

But do you remember the Easterlin Paradox?

At some point, the equation breaks down, and every additional dollar or pound, or each new possession, no longer elicits the same level of satisfaction as it did up to that point of satiation.

However, it's precisely this paradox that our growth-based economic system pretty much fails to take into account. The system does not include the concept of 'enough'. That's why we have reached the point we are at today, where the real goal of economic activity is no longer what it once was: namely, to ensure a better provision of goods and services that people actually need. We have swapped the ends with the means. And, interestingly, although we may not be aware of it as we go through our daily lives, we know precisely what function we and everybody else must fulfil to keep this system producing more growth, and each of us expects everyone else to play their respective part. Or there will be trouble.

Not so?

Then just imagine how the markets would react if Apple stopped bringing out a new iPhone regularly, regardless of whether the new model was really any better than the previous one.

Or what Apple would say if the state took precisely that practice of planned obsolescence as a reason to suddenly increase the tax it levies on smartphones? Imagine the uproar among investors if that led to fewer phones being sold, since investors must always get their returns, at any cost. And what would Apple employees think if their jobs were then cut? Incidentally, such a scenario would also lead to the reduced consumption of new phones.

Companies must constantly bring out new products, consumers must consume those new products, and engineers must come up with new inventions that are thrust onto the market with the help of advertising, while banks must grant loans, and politicians have to create so-called basic economic conditions—which, in reality, means avoiding any measures that might jeopardise the growth of anything that people spend money on. Apparently, growth is the only thing that can guarantee jobs, investment, and tax revenues. Accordingly, everyone within this system must contribute to growth, and everyone is reliant on others doing the same.

This explains why people watch the stock market report with the evening news; they believe it will tell them about economic growth, and therefore about their own future, even if they don't own any shares themselves. As long as the curves on the graphs continue to point upwards, everything appears to be going well. This is despite the fact that those curves say precious little about how well we're doing or not, and tell us almost nothing about our future.

In the old reality of those English fathers of economics, no one ever asked what would sustain all that new growth. And so it appeared at the time to be a perfect upward spiral.

The problem is, it isn't.

As we saw in the chapter on our connection to nature, we humans have not organised our economic activity to be a cycle, but a huge, global conveyor belt on which raw materials and energy are loaded at one end, processed into goods in the middle, and unloaded in the form of money and waste at the other end.

Thus, in the old reality, it was predicted that this form of economic activity would result in 'the greatest happiness of the greatest number'. That's how Jeremy Bentham, another eighteenth-century English thinker, formulated the central concept of utilitarianism. This philosophy promotes an ethical point of view in which the end justifies the means: as long as it brings ever more happiness to ever more people, the economic system is fine. For Bentham himself, writing in his 1789 work *An Introduction to the Principles of Morals and Legislation*, happiness was about maximising positive feelings and minimising negative ones. The economists of his time then created a measure of happiness—or utility—on the basis of monetary value: the value of goods or income was the metric of utility.

As to the question of how the greatest number can partake of this utility, that was explained by Adam Smith in the first chapter of his work *The Wealth of Nations*: 'It is the great multiplication of the productions of all the different arts, in consequence of the division of labour, which occasions, in a well-governed society, that universal opulence which extends itself to the lowest ranks of the people.'[5]

By implication, this means that if the poor are to get a bigger piece of the pie, the pie must continuously increase in size.

Smith's remark about 'a well-governed society' was meant as a dig at the king, who Smith thought should keep out of economic affairs and was overextending his privileges. However, his idea has continued to be cited long after the country stopped being ruled as an absolute monarchy and became a democratic state, which Smith believed bore the responsibility to rein in the power of the big players. The credo today is still that the market is the best organiser of value creation. Disputes—some of them quite vehement—repeatedly arise, over the apportioning of responsibilities between the state and the market, taking the form of arguments over such issues as the merit or otherwise of balancing the national budget, the desirable level of state investments and acceptable levels of state borrowing, or the appropriate use of central bank money.

Since the nineteen-seventies, the greatest influence on economic theories and economic policy-making has been wielded by those economists who advocate affording the private sector as much freedom as possible. Such economists believe that the state should have no role in the economy, since markets are the most efficient distributors of resources, and they provide the best way of balancing supply and demand—which also accelerates growth, resulting in more goods and services to be distributed. This belief has been accompanied by demands for governments to refrain from imposing heavy taxes on the rich, so that they can make investments, create new jobs, and pay higher wages, and thus their profits can trickle down to the lower levels of society.

After a period of too much state regulation, those economists argued, more market force was needed to revive Adam Smith's 'trickle-down effect'.

The phrase 'trickle down' appeared in the US just as much in the speeches of John F. Kennedy as it did in those of Ronald Reagan, as well as featuring in the pronouncements of Margaret Thatcher in the UK. Since the nineteen-eighties, it has served in many countries around the world as the justification for reducing taxes on the richest, and for cutting property and inheritance taxes. It has been used as the reason for privatising state enterprises and for creating the political conditions—by deregulating the financial markets, which means subjecting them to less oversight—for the development of a previously unheard-of range of financial 'products'.

'A rising tide lifts all boats' was the narrative that drove such economic policies.

Nearly fifty years later, we can only conclude that this policy has failed. This is despite impressive figures from the Oxford University–run platform *Our World in Data*, showing that the proportion of people living in poverty has fallen from 94 per cent in 1820 to 10 per cent today. These figures are often quoted, prompting both the former head of Microsoft, Bill Gates, and the psychology professor and bestselling author Steven Pinker to declare the extremely utilitarian view at the annual meeting of the business elite in Davos that no one should be complaining about inequality and the concentration of wealth in the world when the model that led to those phenomena is so effective at reducing global poverty. Pinker, in particular, has been unwilling to take the environmental crisis seriously.

However, Jason Hickel, an anthropologist with a very forensic eye when it comes to analysing data, contests those poverty figures. He concludes that reliable data on global poverty levels have only been collected since 1981. In addition, he points out the controversial World Bank standard applied in those figures, according to which 'extreme poverty' no longer exists anywhere in the world. The standard, set in 2011, which assumes that $1.90 per day is sufficient for spending on healthy food, shelter, and healthcare costs in the US, seems like a pretty bold assumption. If the threshold for poverty is raised to the level enabling what many academics now consider a dignified life, it rises to between $7.40 and $15 a day. At these levels, the success story is exposed as a failure: an assumed poverty threshold of $7.40 a day would mean that no fewer than 4.2 billion people were living below the poverty line in 2019—more than in 1981.[6]

In the same period, global Gross Domestic Product rose from $28.4 trillion to $82.6 trillion. However, only

5 per cent of each additional dollar generated reached the lower 60 per cent of the world's population. And do you know where most of the people whose standard of living has risen above that poverty line since 1981 live?

In China.[7]

If those people are removed from the statistics, little evidence remains of the trickle-down effect from the radical free-market model of economic growth. Not only are many more people living below the poverty line now than in 1981, but the proportion of the growing world population who are poor has stagnated at 60 per cent. Also, the inequality between income and wealth has risen again in industrialised countries since 1980—after about a century of that gap gradually closing.

Taxes on the rich and on corporations are now at their lowest level for decades, and the number of billionaires is rising rapidly. This state of affairs was summarised by Thomas Piketty in his widely acclaimed book *Capital in the Twenty-First Century*, which prompted even market-oriented economists such as Robert Solow to speak of a growing plutocracy. This development has been less pronounced in Europe than in the rest of the world; but in Germany, for example, the indicators for inequality are also rising.[8]

Contrary to expectations, the rich have not spent the tax money they have saved on investing in productive activity, but instead have snapped up many public assets such as infrastructure and buildings. What we call privatisation has led to a situation whereby net private wealth in rich countries has risen from between 200 and 350 per cent of national income in 1970 to between 400 and 700 per cent in 2018, while net public wealth has fallen.[9] That means that this form of growth has led to an increase in the amount of wealth in countries, while the states themselves have become poorer. Previously productive uses of financial capital have been replaced with unproductive uses: costs for the use of assets in the form of rents or leases have risen, although no new value has been created.

Another popular destination for excess capital has been the stock market, where more money could be made with money than by creating jobs. Over the past ten years, 500 of the biggest companies in the US spent $5 trillion buying their own shares, which, for 450 of those firms, meant investing more than half their profits. This was boosted all the more by the tax cuts introduced by the Trump administration: $1 trillion was invested in this way in 2018 alone.[10] Basically, this comes down to fiddling the numbers—the quantity of shares on the market is reduced so that their price will increase. Without any internal changes being made, or their sales or profits increasing, the companies suddenly look like more successful concerns. Of course, the bonuses for those who run these companies are calculated according to this measure of 'success', so they also rise. Two more nicely hockey-stick-shaped curves arise in our new reality, and they show only a small part of the whole.

Poor people, on the other hand, indebted themselves in the run-up to the financial crisis by taking on cheap and toxic mortgages to buy homes that they lost when the housing bubble eventually burst. States then had to bail the banks out, using taxpayer money to prevent them from collapsing. In this way, the profits from this risky gamble were privatised and went into the pockets of the few, while the losses were nationalised and were borne by the many.

It appears that a rising tide actually lifts the yachts far more quickly than it does the rowing boats. And since central banks began flooding the system with cheap money in reaction to the financial crisis, the wealth and income of the top 1 per cent has virtually skyrocketed.

The story of perpetually growing consumption for all has turned out to be a fairy tale, both ecologically and socially. Behind the breathtaking figures, a system has gradually arisen that is destroying our planet, returning ownership structures to those last seen under feudalism, and that relies on constant growth so as not to collapse under the weight of its own inequities.

Despite all assertions to the contrary, the real goal of the current system remains the ceaseless growth of sales, profits, and possessions—whatever the cost.

At the same time, there are some places where these have already grown out of all proportion. I will never forget when, in the summer of 2019, I and a group of others were at the United Nations in New York to discuss the $39 billion lacking every year for the provision of primary education for every child. At the very same time, just 250 metres away, the JPMorgan Chase bank announced it would be paying out 40 billion euros in dividends to its shareholders over the next few months — since it had so many financial resources, it hardly knew what to do with all the money.[11]

Thus it is not more growth to generate enough money to bring far greater happiness to a large number of poor people that is lacking. What is lacking is the economic and political will to re-establish an explicit connection between the accumulation of money and the creation of added value, and to reduce the siphoning-off of unearned income.

What do I mean by that? I mean that we must ask ourselves three important questions when it comes to growth:

How are goods and services produced?

How do they reach consumers?

What happens to the profits generated by these processes?

One thing is certain: there are many actors involved in these processes, all of whom want something in return for their contribution. What happens, however, if all those involved pursue benefits exclusively for themselves, and include only monetary indicators in their calculations? This question is explored by the economist Mariana Mazzucato in her book *The Value of Everything*. She also delves into the history of economic ideas, tracing the way various thinkers through time have explained the creation of added value and wealth.

Until the nineteenth century—so still within the era of Adam Smith and David Ricardo—there was always some kind of objective basis for the determination of how value was added. Value resulted from the various productive combinations of resources such as land and/or materials, or the provision of necessary tools and technical equipment for certain work to be done, or the time it required, or the quality of workmanship. The value of a product or service did not diminish, even if there was no one who was willing or able to pay the price demanded for it. Prices were the result of an exchange transaction in which various interests, power relations, and political conditions all played a part. However, the value of things and services could still be huge for humans' lives, even when they didn't cost anything. Adam Smith illustrated this in *The Wealth of Nations* with what has come to be called the water-diamond paradox:

> Nothing is more useful than water: but it will purchase scarcely anything; scarcely anything can be

> had in exchange for it. A diamond, on the contrary, has scarcely any use-value; but a very great quantity of other goods may frequently be had in exchange for it.

Unproductive activities were also recognised as such, alongside these productive types of labour. Unproductive activities were those in which things that already existed were shunted back and forth — such as in trading, for example, or the distribution of money. Fees could be paid for such activities, but they could not be seen by economic theorists of the days as a productive creation of value. Incidentally, Smith also believed — contrary to the aspirations of financiers — that payments for them should be kept on the small side.[12]

This distinction between value and price was lost under utilitarianism and the mathematisation of the economy. Today, a utility-maximising *Homo economicus* only spends as much money on something as is equal to the added value it brings him. The value of things is thus determined by the price they command on the market, and becomes completely uncoupled from its contents or qualities. Something's price *is* its value. The subjective preferences (of buyers) trump objective resources, and exchange value becomes disconnected from use value.

This has made it possible to create value purely by agreement. And, according to Mazzucato, it has revealed many previously undiscovered sources of unearned income, arising from disproportionate charges levied for the process of shunting existing things back and forth. Do you see what then happens under utilitarianism? That's right: it can become very expensive to organise the creation of wealth in a society for the benefit of the largest number of people.

Mazzucato illustrates this using the example of the pharmaceutical industry. If someone is willing to pay 15,000 euros for a new cancer medication, that becomes its 'value', making it legitimate to charge hospitals the same price for the drug. The fact that this new medication may be very similar to one that is already on the market is deemed irrelevant. As is the fact that people are naturally willing to pay anything not to die. Thus the price is more a reflection of the benefit to a company of being in a position of power, rather than of the creation of a product of intrinsic value.

If you research the jumps in the prices of drugs following pharmaceutical company mergers, you will be amazed by how new company owners sometimes assess the value of the acquired products compared to the predecessor company.

The fact that no new 'value' at all has been created makes no difference to the company's growth indicators, or to GDP. On the contrary, higher overall figures suggest progress and success. So, within the worldview of the exchange-value economy, it remains extremely difficult to argue against such practices.

According to the subjective theory of value, people with high incomes may not only see themselves as particularly successful individuals, but may also claim to have substantially increased the total wealth in society. From a theoretical perspective, however, this is what we call a circular argument. Revenues are justified by the fact that something of value has been produced. However, that value is determined according to ... revenues.

And the circle is complete.

That circle excludes the consideration of such matters as equitable distribution, the most economical way to create value, and the social desirability of the results of value creation.

I am not at all surprised that Mazzucato became known as the economist who 'cancels the business elite's licence to show off', as the German business journal *Manager Magazin* put it.[13] JPMorgan Chase is a good example. Even though the 40 billion euros it paid out in dividends in 2019 came partially from profits it made from speculative high-speed transactions controlled by computer algorithms—which also potentially threatened to destroy whole national economies—paying out so much money gives the impression that value has been created. It also gives the impression that the organisation behind those profits generated them productively. The activities of the financial sector have been included in calculations of GDP since the nineteen-seventies, in parallel with measures to deregulate and reduce oversight over the sector. Its rise has been impressive. Over time, the financial sector's unproductive shunting back and forth of resources in the services of the real economy has become a highly lucrative new business model. Just take a minute to consider again how it works, with expected returns influencing which production processes, remuneration policies, and technological advances prevail in the real economy.

I think we need far greater transparency and public information when it comes to the connections between prices and values.

There needs to be far more intensive discussion of Mazzucato's message, which is that specifically preventing the siphoning-off of unearned value, and adapting accounting practices to make them reflect more objective values, would enable a far more sustainable economy to develop.

Not before time, when this morally blind model of growth is increasing the symptoms of global crises, more attention should be paid to this debate and to the search for progress and good economic policies. Then, hopefully, we will start thinking differently about our concepts and values. And about the way we judge whether and which changes are possible and desirable.

From the product to the process.

From the conveyor belt to the cycle.

From individual parts to the whole system.

From extraction to regeneration.
From competition to cooperation.
From inequity to equilibrium.
From money to value.

Language and the way we use it expresses our goals and aspirations, as well as our concerns. Developing a concept or theory therefore also means demarcating the limits of our thinking, which in turn limits our opportunities for shaping the future. And we shape the future every day with our innovations and technologies, our behaviours and decisions, and the rules we impose on ourselves to enable us to live together. The important thing is the ultimate goals we have in mind when setting those rules.

In a limited world with finite resources, an economic model that relies on constant growth is not sustainable. We must rethink our ideas of what will constitute prosperity for the generation after next. This requires new concepts and fresh ways of thinking to address what will be important in the future. Planetary destruction must no longer be called growth. The pure multiplication of money must no longer be called wealth creation. Limiting growth must become synonymous with overcoming the causes of ecological and social damage.

6

Technology and Progress

'The combined effect of the Industrial and Scientific Revolutions was doubly disruptive, transforming both the structure of society and the way people explained the world.'

Jeremy Lent, entrepreneur and author[1]

The electrification of the world began with the light bulb. Electric light had existed for some time, but it had been a luxury reserved mainly for hotels, offices, and theatres. By the late nineteenth century, wealthy private households were connected to the mains electricity supply, and electric lighting was increasingly being installed in their homes in the form of light bulbs. This was despite the fact that the efficiency of the first generation of bulbs was low, with most of the energy being transformed into heat rather than light. Compared to gas lamps, or indeed candles, however, the light bulb was, of course, a major advance in artificial lighting, and therefore also in the march towards daylight-independent living.

A few decades later, engineers managed to replace the carbon filament, which had been used as the incandescent element until then, with one made of the heavy metal tungsten. It was less quick to burn out than its carbon predecessor, and gave off a brighter light. This was another technological advance, this time in terms of efficiency. Tungsten filament bulbs used only a quarter as much electricity as those with carbon filaments, while producing the same amount of light. This was good news for consumers, but for the electricity plants at the time, it sounded like terrible news.

When the new bulbs went on sale in England in the early twentieth century, the electricity companies feared their businesses would collapse, which seemed plausible at first. They believed that if people could get the same amount of light for less electricity, consumption would go down. Some electricity companies considered raising their prices to make up for the loss.

Interestingly, the exact opposite happened.

The reduction in consumption per light bulb meant there was more electricity on the market, and so its price fell, suddenly making electric lighting affordable for people who had not had the means to pay for it before. A luxury product had become a mass product, and of course that, too, was progress. Paradoxically, however, it meant that it was the development of a light bulb that used less energy than previous designs that led to an overall increase in the demand for electricity. An increase in efficiency—which means simply generating more output from less energy—ultimately resulted in a rise in energy consumption.

Scientists call this the 'rebound effect'.

It is one of the most underestimated obstacles to creating sustainable economies.

If you ask people today what progress means to them, most will think first, and perhaps only, of technological advances. That's not surprising. There has also been a great deal of social progress—for example, legal processes have replaced murder and violence as the normal way people settle disputes; women are no longer burned at the stake for being witches, but can now vote and run for office in most countries and, at least on paper, enjoy the same rights as men; and science is recognised as the method for acquiring and consolidating knowledge and the basis for political decisions—but these seem to be less tangible for people.

In addition, people's opinions about social progress differ more widely when it comes to which developments should be considered good and which should not. An individual's opinions about this are usually connected directly with their identity and often the position they have reached in society. Nonetheless, there is now far greater consensus about the basic ideas of what constitutes a desirable way of living together, and that unity is expressed in documents such as the United Nations' Universal Declaration of Human Rights or the UN's statement of Sustainable Development Goals. However, sometimes there are backward steps, and in many societies there is great disagreement about what we must do next.

Technological progress, by contrast, is seen as *the* success story of human history, tracing a direct line from the hand axe to the smartphone. Every human invention or innovation along that line has expanded the possibilities for human development, and has seemingly confirmed that humanity is on the right path.

Do you remember the chapter on the new reality (chapter two)?

Under 'empty world' conditions, when there were few people with a lot of planet at their disposal, technological progress principally meant the ability to multiply physical strength with fossil-fuel-powered machines—to produce ever more, ever-better goods in ever-shorter times. Mechanical power became the driver of factories, mass production, and therefore also of the growth machine.

The modern concept of development is characterised by precisely that idea of mechanistic and technical progress. The emphasis is placed on the new (in Latin, *modernus*), as opposed to the old. And development always moved, and still always moves, towards expansion: *new* then means *more*, in the sense of bigger, more powerful, and more productive.

Under 'full world' conditions, where the fossil-fuel-based economy threatens to destroy humanity's bases for life, technological progress has been given an extra task: intensification. The new *new* means getting *more from less*, to make it possible to secure and further increase economic growth without destroying the environment in the process. Efficiency gains are now the stated aim, and they are measured not only in monetary terms, but also according to the carbon intensity or resource intensity of the associated growth. That is already a good advance on Robert Solow and his substitutability of natural capital.

This is basically what people mean when they suggest solving global environmental problems—from climate change to species extinction, and the widespread depletion of all our natural systems—with innovation and technological breakthroughs, rather than with state prohibitions and regulations.

Technological progress has helped us exploit nature to generate material growth. Now it is expected to help us stop exploiting nature so much while still generating

GDP growth. Save the planet without having to give up any increase in prosperity; sustainability without sacrifice. On the contrary, it is even profitable, since, even if prices do not even vaguely reflect the ecological truth, it still makes financial sense to keep the consumption of resources at a lower level. This is the idea behind so-called simple decoupling, which enjoyed popularity for a long time because it appeared as if these changes could be made without anyone noticing.

Business as usual—only more efficiently.

Can that work?

The first hint that technological progress alone is not enough is already 150 years old, and goes back to the English economist William Stanley Jevons. He observed that the consumption of coal in England had rapidly escalated in the early nineteenth century despite James Watt's improvements to the steam engine, which had reduced the amount of coal it required by two-thirds compared to previous designs. It was the same effect as would later happen with refinements to the light bulb. A technology was made more resource-efficient, which increased its prevalence, thereby causing an overall increase in consumption that cancelled out or even surpassed the savings from the new, more efficient technology.

'It is wholly a confusion of ideas to suppose that the economical use of fuel is equivalent to a diminished consumption. The very contrary is the truth.'[2] This is how Jevons summed up the effect that would come to be known as the Jevons Paradox, and which we encountered earlier in this chapter as the 'rebound effect'.

Not only do more efficient machines do nothing to slow the consumption of a country's primary resources, but they actually speed it up, due to the spiral of growth between lower prices and the resulting greater prevalence of energy-hungry products. This realisation was of enormous significance for Great Britain, which at the time was almost completely reliant on domestically mined coal to fuel its rapid industrialisation and breathtaking economic rise. Rather than delaying a future energy crisis, more efficient machines brought it closer.

What now?

A century before humanity as a whole came to realise that it is living in a new reality, in which economic activity takes place in a full world rather than an empty one, the British faced the same problem at a national level when it came to coal.

Want to know the solution they came up with?

None—as they were distracted by something else.

And that was?

Oil.

A few decades later, before the extraction of reserves in the British colonies really took off, oil was discovered as an energy source in the US. This marked the start of an age in which energy seemed to be available in unlimited amounts. The rebound effect was forgotten. This additional mineral oil fuelled the idea of perpetual economic growth, which, in turn, appeared to drive perpetual growth in prosperity for all. In most Western societies after the Second World War, if not before, this was not only the general expectation, but also the reality on the whole. It was the model that most non-Western societies aspired to and tried to emulate. In this particular regard, there was no fundamental difference between democracies and dictatorships.

Under those circumstances, it's no surprise that no one bothered about the rebound effect for the next 100 years. But it is also no surprise that that situation has changed dramatically in recent years, as it has become increasingly obvious that the dream of simple decoupling is just a fantasy.

Let's take the motor car as an example.

In the mid-fifties, a normal Volkswagen Beetle required 7.5 litres of petrol to travel 100 kilometres. When VW rebranded the car as the New Beetle and relaunched it in the late nineties, its fuel consumption was almost unchanged. Between the two models lay forty years of technological development, engineering advances, and efficiency optimisation.

Where was all that progress?

It was, of course, in the car. The New Beetle's maximum horsepower was not 30 like its predecessor, but 90 or 115, depending on engine type. Its maximum speed was no longer 110 km/h, but 160. Energy savings that could have contributed to reducing fuel consumption were instead used to increase power. Neither energy nor materials were saved—indeed, the car's weight rose from the Beetle's 739 kilos to the New Beetle's 1,200 kilos.[3]

The automobile is a good example of how multifarious the rebound effect can be. It is one of the products in which practically every efficiency gain has been at least partially cancelled out at some level or another.

The cancelling-out of efficiency gains can also happen as part of the way a car is used—for example, when somebody buys a fuel-saving car and then uses it more than otherwise, taking more daytrips and driving to out-of-town shopping centres instead of walking around the shops in their city centre. Perhaps the proud car owner will now take a better-paid job in a nearby town, and commute to it every day. It can also take place indirectly—for example, if the car owner decides to spend the money she is saving on petrol on something she previously couldn't afford, such as a new smartphone, a city break by plane, or a second fuel-saving car for her partner.

The same is true for producers. For example, they

might invest energy savings they achieve during the manufacturing process in increasing their production capacity, thereby putting more cars on the market. Or they might launch new, more powerful versions of their products, such as the New Beetle. Or they might develop entirely new models, such as an SUV. This has the effect of making the owners of smaller vehicles feel less safe on the roads, or envious of SUV owners' social status or of the significantly greater interior space in such vehicles. More private space inside automatically means less public space outside for pedestrians, cyclists, or other uses, including possible future uses.

Electric vehicles, which are seen as an environmentally friendly alternative, due to their lower emissions of climate-harming carbon dioxide, are also affected by rebound, even when they are charged with green electricity. For one thing, the manufacture of their batteries naturally requires energy, as well as rare earth elements, which are often mined under environmentally harmful conditions; and for another, because developing an infrastructure of charging stations also requires energy and materials.

An electric car such as Audi's e-tron, a giant SUV weighing more than 2.5 tonnes and requiring a battery weighing 700 kilos just to start moving, shows that technology can perhaps make things go faster or work more efficiently, but that this is not necessarily a good thing. Six to ten tonnes of carbon dioxide emissions are generated each time by the manufacture of the 100 kw/h batteries that large electric cars depend on.[4] An economical petrol or diesel car with an average mileage would have to be on the road for four years to match those emissions.

Concentrating on just one single product, or one single aspect of a product, means failing to see the bigger picture. That picture includes changes that can occur in the overall system into which newly invented technologies are introduced. This is what scientists call systemic thinking. According to this approach, the solution would be to have fewer cars, and to make those that are manufactured to be as environmentally friendly as possible. You might recognise this from the earlier chapter on nature: when one element is altered, the entire dynamics of the processes in which that element is embedded also change—and likely, also, so does the quality of the other elements it is connected to.

It is for this reason that academics in transformation research speak of socio-technological and socio-ecological systems. New technologies do not leave their environment or the people in it unaffected. Systemic thinking enables us to see that technological change can affect many aspects of our lives, including what we know and how we communicate that knowledge; our work routines and daily habits; and even vested interests, power, property, infrastructure, and the way we shape the landscape around us. Conversely, those systemic structures have an influence on which technologies are the next to be considered useful, interesting, or desirable, or have a good chance of gaining currency.

But why is this important?

Our industrialised society has made huge advances in improving efficiency over the past thirty years. The German economy currently consumes less energy, emits less carbon dioxide, and uses fewer materials, and therefore fewer natural resources, to produce one unit of Gross Domestic Product than ever before. Unfortunately, as described earlier, we are still very far from only using our fair share of natural resources. Do you remember the concept of Earth Overshoot Day? The date on which our consumption exceeds the amount that the Earth can naturally regenerate is being reached a few weeks earlier every year. Just to reiterate: that date fell in May for Germany in 2019.

Despite all this—I'm sure you know where I'm going with this—the goal of continued growth by means of innovation and progress is never questioned. On the contrary, returns on investments, profits, sales, and economic growth are taken as the main indicators of successful innovations.

Ultimately, as long as the economy is expected to grow continuously, the goal will never be to live well but within our planet's limits, and to seek solutions for doing so—which would correspond to the concept of the systemic efficiency of ecosystems. The idea of efficiency in current economic science is, to put it bluntly, the philosophy of 'two for the price of one'.

We have also seen that concepts have the ability to affect reality in tangible ways. 'Two for the price of one' is a very real way of quashing the dream of absolute decoupling—that is, of really halting the increase in the consumption of resources—before it even takes root.[5] We continue to push the curves for carbon dioxide emissions and resource consumption highly efficiently and ever upwards.

And we do the same in many areas of our lives.

We heat our homes more efficiently than ever before, for example, and our buildings are thermally insulated, but because we constantly expect more dwelling space per capita—that is, individuals all demand ever more living space for themselves—our energy consumption for heating has not gone down.

The same goes for electrical appliances. They now use less power, but we also own more of them. And they often have much shorter lifespans than before.

What used to be considered luxuries are now seen as normal, and their supply must be secured at all times and in all situations, while owning or engaging in them is seen as a measure of what constitutes a good life:

- A constant, direct supply of running hot water;
- A car for every family;
- A washing machine in every home;

- A flatscreen in every living room and bedroom;
- A car for every person;
- Fresh strawberries even in winter;
- Air-freighted mangoes processed into bite-size portions; and
- A trip by plane every second weekend.

One of the starkest examples of humanity's problem with putting a stop to this expansion is so-called geoengineering—the term used to describe attempts to slow down climate change artificially. Large-scale reforestation projects or the rehydration of wetlands to bind carbon dioxide are often quoted as geoengineering measures. However, since the land that such projects would require in order to be of any significance to climate change is in competition with the land requirements of expanding human settlements, infrastructure, and agriculture, technological solutions are often invoked. Such suggestions include, for example, deploying huge folding mirrors in space to shade the Earth from the Sun's heat, or using planes to dump tonnes of sulphur into the atmosphere to reflect sunlight away, in an effect similar to that caused by some volcanic eruptions. Some scientists are also considering using fertiliser to deliberately trigger algal blooms in the oceans, or grinding down mountains and scattering the dust, since carbon dioxide is fixed when mountains are weathered.

You think all this sounds like something from a James Bond film? Then you should know that almost all climate models that show us as still being able to reduce global warming by more than two degrees are based on the assumption that geoengineering measures will be used in the foreseeable future. Without them, the models fail to produce that reduction in warming.

The only problem is, those technologies are not yet available in any workable form. Either they have not yet been adequately tested, or have proven to be dangerous, or they only work on a small scale, without any evidence that they can be scaled up unproblematically. In fact, the effects they are expected to have are already included in the prognoses as 'negative emissions'.

And since we have such confidence in technology to solve these problems, geopolitical tussling has already begun over the raw materials and oil deposits freed up by the melting ice in the Arctic. However, the question is not how much coal, oil, and gas is still hidden underground, but how we deal with the problem that the Earth's atmosphere cannot absorb the resultant carbon dioxide without it ruining our currently human-friendly climate.

Also, our appetite for more and more is growing so rapidly that we are unable to develop renewable sources quickly enough to satisfy it. The fact that renewables are now often cheaper (from a market-price perspective) than coal-based energy is great news. It will have a huge impact on long-term investment decisions when it comes to energy production. But this did not stop the oil giant Aramco from becoming the world's most highly valued company shortly after its stock market floatation in 2019. The world's hunger for energy is so great that renewables are not seen as a replacement for fossil fuels, but as a supplement to them. Just as oil was a supplement to coal a century ago.

In other words, if we continue to deploy technological advances as we have so far, without clearly affording them a function other than that of enabling short-term economic growth and the continued increase in consumption, all we are doing is callously continuing to leave the problems to the future to solve.

To be honest, my jaw dropped when Elon Musk presented his Cybertruck in late 2019—with its 'almost impenetrable exoskeleton' and its 'ultra-hard 30X cold-rolled stainless-steel structural skin' and its 'Tesla [armoured] glass'—as the next-generation sports car. It boasts of insanely rapid acceleration and a payload capacity of 1.7 tonnes, while the Model S was to weigh in at 2.1 tonnes. When and where are those features needed by any vehicles, outside of motor races and motocross rallies? Excuse me, but what is the environmental footprint of such a thing? Still, according to Musk, 250,000 such vehicles were quickly pre-ordered in the US, although the release has since been pushed back to 2023.[6]

What function might this truck fulfil, beyond its barely useable features?

As we have already seen, this question is asked astonishingly rarely. When it is explicitly investigated, as the sociologist Philipp Staab does in his book *Falsche Versprechen. Wachstum im digitalen Kapitalismus* (*False Promises: growth in the age of digital capitalism*), a different kind of decoupling becomes apparent: that of technological progress from social progress. Once again, we see that increased sales and growth are the overriding goal, and innovation is seen as a means of achieving that:

> In societies with material overabundance, products are rarely bought only for their utility value, but most often for their distinction potential, that is, for their ability to set the buyer apart symbolically from others through conspicuous ownership of scarce products or those which carry certain social connotations. Economically speaking, this is advantageous because it means the needs of consumers are basically infinite and independent of products' useful value.[7]

Humans are excellent problem-solvers, but when the problems are not properly defined, progress tends to pass them by. At this point, if not before, the example of the Cybertruck gives me a strange feeling, and I am extremely glad that Germany's national technical safety inspectorate considers it ineligible for approval. On a symbolic level, this vehicle is like something out of a *Mad Max* movie: I don't care who or what gets in my way, from tomorrow I'll simply mow it down in my nicely armoured tank powered by the full force of the Sun.

If this is the choice, I prefer modern trends such as mindfulness, yoga, country walks, forest bathing, self-actualisation, digital detoxing, and time prosperity. But did you know that the rise of such trends was predicted as the natural development of a society that has grown rich on technological progress?

And can you guess who made those predictions?

Economists.

Ninety years ago, John Maynard Keynes, one of the most important economists in history, wrote an essay entitled *Economic Possibilities for Our Grandchildren*. In that work, he considers what will become of humanity once its economic problem is solved, meaning when its material needs are met and everyone has enough to cover their requirements. In view of the ever-increasing productivity he saw, Keynes assumed that point would be reached in 2030. Then, he predicted, we would all need to work only fifteen hours a week to secure our supply of everything we need. At that point, growth would stabilise at an appropriate level, and the economy would simply continue to operate at that rate.

So the question John Maynard Keynes asked was:

What will we do with all that new free time?

Indeed, what do we do with it?

Can you guess?

Of course, we enjoy life.

That's what Keynes believed, too. His idea was that we would use the time to tend to our personal wellbeing and reach our full potential as human beings. Spend time with friends and family, educate ourselves, and devote ourselves to art and culture.

Astonishingly enough, Silicon Valley is also currently thinking once again about the development of human potential. Only, unfortunately, it isn't coming to the conclusion Keynes hoped for.

The internet, having started out as an amazing new form of communicating, connecting, and exchanging knowledge and information, has become the embodiment of what the economist and architect Georg Franck called 'the economy of attention' at the end of the last century. 'Attention from other people is the most irresistible of drugs', he wrote, describing attention as a kind of currency that is in limited supply.[8]

Digital service-developers became extremely interested in this analysis, and began designing all their products and standard settings with the aim of persuading people to spend as much time as possible with their content. Their success is measured in visits, clicks, and likes. However, as we know, behind the 'free' content lurks a lucrative business trading in personal information and advertising revenue: bespoke offers are tailored to our every mood, and made so quickly and easily available that all our purchasing inhibitions are subverted. With every mouse click, what began as a digital currency is converted into actual monetary currency. And although the information is produced exclusively by users, it is the companies that make money out of it.

In the opinion of some of those who have now left the digital industry, this kind of technological progress has led to situations where we must talk of changes to human attention activity and learning behaviour, as well as to our social relations and the way we converse with one another. We are no longer only worried about employment. We must also consider the state of our democracy, our social communications, and the manipulation of human beings.

Tristan Harris, a former design ethicist at Google and founder of the Time Well Spent initiative, which later became the Center for Humane Technology, was searching for an overarching term to describe the dark side of the digital revolution. He wanted to find an analogue to the phrase 'ecological degradation'—used to describe such phenomena as climate change, biodiversity loss, water scarcity, and desertification as part of the same basic pattern.

In talkshow appearances and interviews, Harris describes this pattern as 'human downgrading'. It's the downgrading of our attention, of our sense of proportionality in our behaviour, of the democratic processes of communication, and of our social relationships, all the way to our social media addictions.[9]

Harris's idea of 'human downgrading' makes one thing clear to us above all: technological progress simply for the sake of it, or for the sake of raking in as much profit as possible, rarely pays much regard to the system in which it is embedded.

Is that an anti-innovation statement?

Empirically speaking, not at all, since limiting the overexploitation of people and nature could drive precisely the innovation agenda that our new reality requires. People can often be pushed to new heights of creativity when conditions are constrained. They overflow with ideas about how to deal with the limited resources available—which is entirely in line with Darwin's observations about evolution in limited ecosystems.

So the positive news is that technological progress is neither intrinsically good nor evil. It can and will be very important in the transformation of our nature-forsaking conveyor-belt economy into a circular economy. Transitioning to a comprehensive renewable-energy system and a sustainable transportation structure will also require the development of new technologies. But for this to take place, we will need to completely rethink our ideas about what constitutes desirable progress towards achieving those goals, and away from the mere multiplication of money. If we do not, we will merrily continue confusing the ends and the means, staring in a stupor at the speedometer, completely oblivious to where we are heading and how much fuel is left in the tank.

Technological progress is considered the most visible sign of human advancement. However, as long as we continue to ignore the issue of how technology is embedded in the environment and in society, we will fail to see where we are heading. In order to live well together in our new reality, we must also change our idea of progress; otherwise, we will simply be putting off solving the problems to the future.

7
Consumption

'Too many people spend money they haven't earned, to buy things they don't want, to impress people they don't like.'

Will Rogers, 1879–1935, American comedian[1]

One of the bestselling non-fiction books of recent years is *The Life-Changing Magic of Tidying Up*. Its Japanese author, Marie Kondo, topped the sales charts in her home country for a long time. Her decluttering books have now been translated into forty languages, with global sales of more than seven million copies—predominantly in Western industrialised nations. Apparently, that is where people need to be told how to tidy up properly. Which is understandable, as you can only have a problem with clutter if you have acquired far too much stuff in the first place.

Kondo's method is based on the simple idea that

you can't create true order among your possessions if you possess too much. That concept is particularly apposite in a country like Japan, where living space is so expensive that spreading your stuff out over a greater area is simply not an option.

So the author recommends creating piles of things according to categories—she never tidies up room by room—such as clothes, books, papers, odds-and-ends, and mementos, for example, then sorting them according to how you feel about them, asking, 'Does this possession spark joy when I hold it in my hand?'

If not, get rid of it.

It is not only through books that Kondo teaches her tidying method. She now also gives training courses for tidying consultants, and a while ago even had her own multi-episode documentary series on the streaming service Netflix, in which she helped overwhelmed Americans clear up their cluttered closets, kitchens, spare rooms, and garages packed with the products of prosperity. Incidentally, the people she helps throw things away in the series are not compulsive hoarders. At the end of the process, they all appear incredibly relieved as the garbage truck drives away, piled high with plastic rubbish sacks.

Do you remember the Easterlin Paradox? The realisation that from a certain level of prosperity, extra wealth does not mean additional happiness?

In a manner of speaking, Kondo made the movie.

My first thought as a sustainability researcher, of course, never even came up in the series: What if those people hadn't bought all that stuff in the first place? What if it had never been manufactured? We would have far fewer piles of plastic sacks filled with waste.

Two suggestions generally come up when we talk about ways to rethink our economic system so as to make it sustainable and within the ecological limits of the planet. The first is now familiar to you as simple decoupling, which involves relying on innovation and technological advances to reduce the depletion of nature, without the need to sacrifice any of our prosperity. Unsurprisingly, it is the more popular of the two proposals; but, as the rebound effect shows, we humans have so far failed to achieve our aim using that strategy. We have seen the same pattern of rebound in the use of human resources such as time, attention, and money.

Of course, it is not only the supply side that must play an important part in this endeavour, but also the demand side—consumers—and that's the starting point for the second proposal for achieving a sustainable economic system. If increasing economic growth is unable to preserve nature in its present state, let alone to allow it to recover, then material prosperity must decrease. Of course, that's the less popular option, as it means living on less, or, in other words, making sacrifices.

As we saw in the chapter on nature, the environmental damage caused by the manufacture and use of a product is not priced into any economic balance sheet or calculation. This means that what we pay for a product does not reflect its actual price. In principle, that is an exercise in cooking the books, and is often cited as such by those who criticise the metric of Gross Domestic Product. Nonetheless, this type of accounting was and is a tried-and-tested way of making products artificially cheap. It shifts the burden created by the production or consumption of something onto others who cannot defend themselves, because they have no voice and no power.

Let's take the example of a flight from Frankfurt to New York.

Depending on the time of travel, a return ticket for such a flight might cost less than 300 euros. As well as all the other costs, that price, of course, also includes the fuel necessary to fly the passengers there and back. However, the ticket price does not include the cost of removing the carbon dioxide emitted by the flight from the Earth's atmosphere. Airlines do not add that cost to the price of tickets, just as fuel suppliers do not add it to the price of the kerosene they sell to the airlines.

Everyone, including the passengers, simply assumes that the Earth's atmosphere will just absorb the extra 3.5 tonnes of carbon dioxide emitted per passenger by the flight.

Hence, 'external costs', a phrase beloved of economists, is a totally crazy term. External to what, exactly?

Well, apparently, external to the category of things we feel responsible for. Although we have long used the atmosphere as a rubbish tip in any number of ways, and dumped our greenhouse gases into it, we vehemently deny any responsibility for cleaning it up. Ultimately, that price is paid by small island states as they sink into the sea. Or poorer people who cannot afford to adapt to climate change; for example, those who cannot afford to re-sow their crops or to rebuild their houses after they are destroyed by storms, or to move to other areas after their home regions are flooded. We are also burdening our children and grandchildren with that responsibility, as they will have to live in the world we leave behind for them.

Social scientists call such denial of responsibility 'externalisation'.

In his book *Living Well at Others' Expense*, the German sociologist Stephan Lessenich shows how Western prosperity is partly built on an offloading of the real costs to others. And we passively or deliberately ignore that fact so we can continue to prosper. That is the pseudo-reality I described at the beginning of this book. Lessenich's term for it is 'the externalisation society'.

'We are not living beyond our means,' he writes. 'But at the expense of others.'[2]

German farmers fatten their livestock on soya that isn't grown in Germany. It's imported from South America, where rainforests and grasslands are destroyed to cultivate it on a massive scale. Germany, in turn, produces far more meat than it consumes—so it exports the surplus to countries where local farmers who don't feed their livestock with cheap soya cannot sell their own meat as cheaply. The financial benefits achieved by causing damage in one location lead directly to more damage in a different, distant location. Cause and effect are decoupled from each other and spread across the entire globe.

Another example is biofuel, which a few years ago was Europe's great hope for improving the climate footprint of its transport sector. The carbon dioxide emitted when biofuel is burned can be reabsorbed by the plants from which the next batch is manufactured. So, theoretically, that should be a sustainable cycle. However, because the quantity of fuel consumed in Europe far exceeded the amount of canola or sunflowers that could be grown on the available land, biofuel had to be imported from other parts of the world. I think you can guess what's coming: rainforests were cleared in South-East Asia to make way for palm oil plantations to cover Europe's demand for plant-based energy. The fact that the fires set to clear those forests released huge amounts of carbon dioxide that was previously fixed in the forests and soil is something we prefer to, well … externalise.

Luckily for us, we don't really get wind of all that. In Germany, we're proud of the figures that show the country has stable or even growing expanses of forest cover. However, what we don't talk about is the detrimental effect that all those monocultured hectares have on biodiversity. As the recent very hot summers have shown us, they also lack the resilience to respond to climate change. And yet we're constantly told that people in poorer countries must learn to treat the environment better.

Interestingly enough, economists also see the answer to this in economic growth. The Kuznets curve is named after Simon Smith Kuznets, an economist who lived in the US. It describes the hypothesis that, as an economy grows, market forces first increase and then decrease economic inequality. The curve's development is impressive: at first, people all have a similar amount, then a few become rich, and finally there are riches for almost everyone.

The trickle-down theory that we met previously in connection with economic growth has been recast to apply to environmental sustainability. Can you guess how? That's right. In this context, the theory says that the amount of environmental pollution sinks as per-capita income increases.

Or, to put it another way: the richer societies get, the more interest they have in maintaining a clean environment, and the more economic means they have to create and maintain the necessary infrastructure for that.

Oh, really?

If you look at the issue of waste sorting and recycling in Germany alone, it appears, at first glance, that a great deal of prosperity is required to afford what is ostensibly one of the most efficient recycling infrastructures in the world. Estimates put the cost to German consumers at around 1 billion euros a year. And that doesn't even count the time required to participate in the system as conscientiously as people in Germany do.

So, is it still a model for success?

Well ... first of all, in 2020 Germany generated more waste per capita than almost any other country in Europe, with only Denmark, Luxembourg, and Malta producing more. But all the waste from a country does not stay in that country. Waste management is an export industry in Germany. One study carried out by the University of Würzburg-Schweinfurt found that Germany exported more waste to other countries than machines (by tonnage) in 2018. One-fifth of its plastic waste is sent abroad, mostly to Asian countries such as Malaysia, India, and Vietnam, where some of it is recycled, but the rest ends up on tips, in rivers, or in the ocean. Every day, 175 old TV sets arrive in Africa from Germany, in countries such Ghana, Nigeria, and Cameroon, where they are stripped down, and any unsaleable parts end up on rubbish tips.[3]

As you can see, Germany's lifestyle is not more environmentally friendly simply because it is more prosperous. Quite the opposite, in fact. It does protect its own environment with more stringent regulations, and runs a relatively well-developed waste-disposal system, compared to most other countries, but its people are woefully uninterested in whether the overall balance is positive. The country outsources what it finds unpleasant, and buys in what it needs.

The same is true across Europe. Of all the continents, Europe is the most dependent on using land areas in other parts of the world. Until the UK left the European Union, Germany required a so-called land footprint of no less than 640 million hectares to maintain its lifestyle. That's around one-and-a-half times the size of the pre-Brexit EU with twenty-eight members. The total went down by around 80 million hectares when the UK was removed from it. Germany's figure is about the same as the UK's.[4] The importers who buy products based on exploiting that much land are generally more interested in low prices than whether the land can continue to produce such yields sustainably. Germany's wealth and prosperity—in other words, its privileged market position—gives it the power to do this.

All this is what is meant by externalisation.

At the national level, the Kuznets curve does apply, at least in the sense that local levels of air and water pollution do tend to fall when a population gets wealthier. (Notwithstanding the cases where car-exhaust emissions data is tampered with.) On a global level—and the majority of the environmental problems we have to deal with today are global in scale—this equation of wealth and environmental protection obfuscates the issue. So we have no choice but to aim directly for reductions in environmental consumption and to demand budgets that help us achieve this. But then this is immediately decried as imposing a list of 'nothing but do's-and-don'ts and doing-without'. 'Doing without': the phrase alone is enough to get many people's hackles up.

But what does 'doing without' things mean in real terms?

We can only voluntarily do without something if we're entitled to it under the prevailing rules. But, under the rules of sustainability, the prosperity that Western countries enjoy and that developing nations aspire to should never have come about in the first place.

In view of this, 'doing without' in rich countries—with

our conspicuous consumption of tank-like trucks and advice books to help us muster the courage to throw our things away—must mean nothing more and nothing less than 'doing without' the ruination of our planet, thereby preserving the basis for life in the future.

That is a big ask, of course.

Can't it be a bit smaller in scale?

No, I'm afraid it can't.

Let's turn the question around and ask what things, for example, we absolutely can't do without that would still leave us able to consider ourselves well provided for.

In fact, the term 'supply security' describes what is needed so that everything that is necessary to cover people's basic needs—food, clean drinking water, shelter, energy, healthcare, and education—is permanently and reliably guaranteed. As we have seen, expectations of what counts as 'necessary basic needs' rose steadily over the past century, but it really began to skyrocket a few decades ago. In the white heat of technological progress and amid economic indicators that are blind to nature, we have completely lost view of the fact that there is such a thing as the supply paradox: when parents constantly strive to give their children a better life than they had themselves, and consistently confuse 'having a *better* life' with 'having *more* possessions', the eventual result is that all children have a less good life. On a planet with limited resources and an increasing population, supply security cannot be taken to mean ever-increasing consumption.

When those who oppose doing without ask what they would get in return for their sacrifice, what would ease the pain of the loss they would suffer, the answer must be that we are investing in peace and supply security for the generation after next. After all, imagine if the countries of Africa, Latin America, and Asia decided at some point to 'do without' exporting their resources and the products of their land to us, and instead use them themselves.

A first step towards resolving the supply paradox would be to correct our accounts—which would also correct price structures. The price of many products would then increase to reflect the true costs of their production, transportation, and disposal after use. Carbon pricing is an attempt move in that direction. It aims not only to influence consumer decisions, but also to create a cost advantage for innovations that aid the development of carbon dioxide–free products. Or, conversely, it makes the environmental damage caused by a product visible in its price—which also takes us a step closer to a more objective concept of value creation. The digital revolution could aid in this process for once: carbon dioxide trackers or digital markers for individual raw materials and product components would be an excellent advance, giving markets a better chance of guaranteeing long-term supply security.

It seems that keeping a level head in the debate between 'more' and 'less' is not easy. We are, after all, used to having ever more things at our disposal. The best symbol of this is the smartphone: music, movies, knowledge, contacts, consumer goods—all available via a single device with 120 million times more processing power than the computer on board Apollo 11, the spacecraft that landed men on the Moon fifty years ago.

In one of his lectures, the sociologist Hartmut Rosa called this 'the constant compulsion toward expanding one's own share of the world'.[5] Our society is set up in such a way that the present is constantly trying to trump the past. There is a relentless drive for more—not just in technology and business, but also in the social, and even spatial, sense. Every fashion, every job, every pleasure, every vacation might be yesterday's news by tomorrow. And the attention economy of constant advertising, news feeds, self-promotion, and information can be relied on to help that expiry date come closer ever more quickly.

This is compounded by the fact that we not only have ever more things and opportunities at our disposal, but that those things are also available in an ever-increasing number of variations. This simply overwhelms us, as two American psychologists proved in a field test a few years ago. They set up two stalls outside a California deli, offering tastes of different kinds of jam that were for sale. One stall had six different varieties; the other, twenty-four. Unsurprisingly, the stall with the larger selection of jams attracted more customers, but in the end, customers bought far fewer jars from that stall than from the one offering only six kinds of jam. Customers had less choice at that stall, but they obviously also had less difficulty in reaching a purchasing decision. The pleasure of choosing does not automatically increase as the number of options increases. The psychologist Barry Schwartz named this phenomenon 'the paradox of choice'.[6]

But it gets more complicated than that.

Just ask yourself honestly whether your quality of life would really suffer if you had to do without a little consumer choice and make fewer purchases. Thankfully, there are now numerous other studies on this, and they all come up with the same clear message: having more and more does not mean more and more improvement. Having more and more satisfies something in us, but it also fuels a concern.

The conveyor belt that transforms the environment into prosperity is not only driven by our desire for ever more. It is also kept in motion by our fear of having less. This fear—of having less than our ancestors, less than our neighbours, less than people we aspire to be like—makes it difficult to share and to do without things. And the more our culture equates the idea of a successful life and career with having more and more possessions—and, more importantly, with having more possessions than others—the faster the conveyor belt runs.

The American psychologist Tim Kasser investigated the social consequences of the economisation of culture.[7] He wanted to know how our materialistic values affect our wellbeing and self-esteem, and found that materialism is both an expression and a cause of insecurity and unhappiness. This is because it primarily addresses extrinsic goals—that is, those originating from outside a person—and the desire for affirmation from others. The price of something, or the amount of attention it attracts (such as fame, likes, and clicks), is seen by us as a reflection of our intrinsic value. Just as Mariana Mazzucato ascertained in the case of goods and services, this theory of value disregards any feeling for what constitutes a valuable member of society. And if we lose our important job or big house, or if our followers suddenly decide they don't like us, our sense of self-worth is in danger.

Little wonder, then, that Kasser discovered that the more materialistic a person is, the more likely they are to feel stressed and insecure, as well as more inclined to depression.

The same conclusion was reached by the lawyer,

educator, and long-time president of Harvard University Derek Bok in his meta-analysis of political recommendations arising from happiness research: 'The findings of psychologists convey a warning that being preoccupied with getting rich carries a substantial risk of leaving one unhappy and disappointed in the end.'[8]

What would Jeremy Bentham have to say about that?

He would be tearing his hair out. By reducing utilitarianism to nothing but constantly increasing consumption, economists declared a growth narrative to be natural and eternal, which in fact does not make us ever-increasingly happy at all.

Ever-increasing happiness?

In fact, that's an impossibility.

People are not mechanical systems, but biological ones. Our brains are constantly adapting to prevailing conditions. Being flooded with too many happiness hormones at once would be more than our brain could deal with, nor can it continue to work at a maximum level of performance over an extended period of time. Living systems such as human beings and the natural world need to be allowed time to regenerate if they are to flourish. That's one reason why happiness researchers don't measure things by constantly rising hockey-stick curves, but in scales of one to ten.

Despite this, our incentive systems, organisational structures, political programs, financial markets, and economic indicators are all designed with one goal in mind: more. As a result, it is becoming increasingly difficult to overcome the conditions that provide for this very particular kind of unhappiness.

This is the second of Kasser's discoveries:

The relationship between material, social, and/or environmental values vis-à-vis materialistic values is like a seesaw. When one side goes up, the other goes down. When the *Homo economicus* perspective is culturally and structurally dominant, everything is about status, power, and money. At the same time, compassion, generosity, and environmental awareness, as well as questions about what is enough and concern for the wellbeing of everyone as a whole are erased from theories and worldviews. And when the concept of 'I' includes ever less 'we', the whole of society has a problem. However, the good news to emerge from Kasser's research is that this seesaw of values also works in the other direction. As soon as more currency is given to social and environmental values, the importance of material values decreases. And the conveyor belt slows down. This has led the behavioural economist Armin Falk at the University of Bonn to propose a categorical imperative for the age of climate change: consume the way you *would like* everyone else to consume.[9]

Now, suddenly, it all seems pretty simple, right?

However, buying less impacts negatively on producers' sales. As you may remember, that results in a recession. So we are exhorted to do our duty not only as good consumers, but also as good citizens. We need a change in politics and policy-making so that sustainable consumption, production, and investments are no longer seen just as possible by-products of an economic-growth agenda, but are considered its aim. You want a short and catchy formula for that? You've already heard it: growth must be the means, not the absolute end.

We need to break free of the Easterlin Paradox, the Jevons Paradox, and the supply paradox, and enter into a new social contract for a high quality of life with a low ecological footprint. That is possible.

Our consumer behaviour in the rich West is only possible because we externalise its costs. Marking our self-esteem in terms of possessions and status does not bring us happiness. Changing how we consume and redefining the role of consumption in our society is therefore an important key to sustainability. Reconciling social and economic goals should form the central focus of this re-evaluation process.

8

The Market, the State, and the Common Good

'Complexity economics shows that the economy, like a garden, is never in perfect balance or stasis and is always both growing and shrinking. And like an untended garden, an economy left entirely to itself tends toward unhealthy imbalances.'

Eric Liu and Nick Hanauer, economists[1]

Ithaca is a small American town in upstate New York best known for its university, which has produced a great number of Nobel Prize winners. Until well into the nineteen-fifties, the railway was the most reliable and economical way to reach the city. Of course, there were roads for cars and buses, and Ithaca had its own airport, even back then. The trains ran all year round, through all kinds of weather. But from the middle of the last century, increasing numbers of people were able to afford a car of their own, so they only resorted to the train when ice and snow meant they couldn't use the roads. At the end of the nineteen-fifties, the railway company closed the route for passengers. It was no longer economical to run.

A few years later, Alfred E. Kahn, an economist teaching at the university in Ithaca, wrote an essay about the fate of the railway line. The title of that paper has become a dictum for all processes whose results are neither the original intention nor the ideal outcome: 'the tyranny of small decisions'.

From their own perspective, all the commuters who turned to cars, buses, or planes to reach Ithaca rather than the train acted reasonably and in their own individual interest. Overall, however, their actions supported a process that culminated in the cancellation of the railway service. Many individual decisions that were all reasonable in their own right combined to produce a result that no one would have actively chosen.

How could that happen? How was it possible that the free market, which was always supposed to result in the greatest benefit for all—as long as everyone made rational decisions to maximise their own benefit—had produced a result that was disadvantageous for all, precisely because everyone had thought only of themselves.

Could there actually be such a thing as market failure?

At its core, the question goes like this: As long as producers have the freedom to produce what they want, and consumers have the freedom to consume what they want, then surely the goods that society wants will be produced and distributed? Surely the market acts as a kind of central coordinator? So why didn't it work out here? It's no wonder that the precise role of the state in relation to the market was, and still is, a matter of much discussion.

With the collapse of communism, the great questions about how to create justice, equality, participation, and progress among people, and how a state must be organised to achieve this, seemed to have been answered. Politics had fulfilled its role as a guiding hand, and the state became a mere night-watchman—responsible for security. The American political scientist Francis Fukuyama even spoke of 'the end of history'. The Washington Consensus on the advantages of a globalised world economy was formulated in 1990, and the World Trade Organization (WTO) was established in 1994.

Some years later, in 2003, I was in Cancún, Mexico, standing with hundreds of other demonstrators behind a fence on the other side of which ministers from 146 countries were gathered at a WTO conference. They were there to discuss, among other things, the consequences of globalised trade in agricultural goods. I was there as a volunteer for the German Federation for the Environment and Nature Conservation to protest against this policy—because, like many others, I could see that globalisation as it was being pursued was having a huge adverse impact on the environment. The main problem, however, was that globalisation benefitted the big corporations in the global North, while the small producers had to pay the price. And that price was much steeper for the small producers in the global South than for the subsidised farmers of the North. On the day that the biggest demonstration against the conference was planned, a protester climbed onto the security fence just a few metres from me and stabbed himself in the chest with a knife, in full view of everyone. He died shortly afterwards in hospital.

We were all in shock.

I later learned that the man's name was Lee Kyung-hae and that he was fifty-six years old. He was a farmer from South Korea, and was seen as something of a guru of sustainable agriculture. He ran a model farm, mostly keeping cattle, where he taught students low-impact, natural livestock-farming methods, until the

South Korean government opened its borders to beef imports, and thrust cheap, factory-farmed Australian meat onto the market. Lee was unable to compete with that produce. He lost his farm and his land to the bank, and he was not the only South Korean farmer to face such a fate. Lee had often tried to publicise the farmers' story, and had now travelled to Mexico to grasp what he saw as his last chance to draw the public's attention to the effects of globalisation.

What had happened here?

What does it tell us about our modern life and the interactions between the state, the market, and public interest?

In the thirty years following the collapse of the Soviet Union, the world changed on a scale never seen before. Many national regulations were scrapped, and new international security mechanisms for investments and transactions were established, all with the aim of achieving globalisation. Worldwide value-creation chains arose, which came to be administered by an ever-smaller group of huge corporations. For example, five corporations now account for 70 per cent of both the export and import of agricultural commodities.[2] The market value of those companies is greater than the Gross Domestic Product of many entire countries.[3] At the top of the list are digital companies, which can move their administrative headquarters particularly easily to places with more favourable conditions—which means not only favourable infrastructure, but also and more importantly, lower taxes and higher state subsidies. As we saw in chapter four ('Humans and Behaviour'), competitiveness—originally a way of comparing companies with each other—has become a standard for nations. Companies that can afford to do so are able to compare government-imposed labour standards, social security contributions, legal regulations, and environmental laws from one end of the world to the other. There are now even law companies that specialise in suing state governments for their environmental and social policies on behalf of companies whose profits do not live up to the predictions on which their investment decisions were based.

Oligopolists—leading companies that dominate areas of the market, together with a small number of competitors—work internationally, while states are forced to act nationally to protect their home companies, because no state can afford to let those businesses collapse. They are 'too big to fail'. We last saw this during the financial crisis of 2008, when the big banks had to be bailed out with hundreds of millions of dollars of taxpayers' money, because their collapse would have sent the global financial system into turmoil and perhaps a state of collapse. The tyranny of small decisions became the tyranny of the big players.

How could the state, which should ideally represent the interests of its citizens and of the common good, be put on the defensive in such a way?

When it comes to the way that markets balance supply and demand, today's basic economic models recognise only two actors: producers and/or companies, on the one hand, and consumers and/or households, on the other. The state does not feature at all—or, at most, it appears as a customer. This is despite the fact that the regulations and incentives the state can and does use to manage the production of goods and services have at least as much influence on supply as demand does. Astonishingly enough, this reduced view is what shapes the current political debate over who is entitled to intervene, and in what way. Three objections, which are only assumptions, or perhaps even prejudices, are particularly prominent in this debate.

They are:

- State regulation (also known as regulatory policy-making) inhibits innovation and therefore also progress;
- The market and companies always have better solutions than the state, and must therefore not be restricted in their actions; and
- Prohibitions restrict the freedom of market

participants, and in particular in this case, those of consumers.

Let's examine those assumptions, one by one.

In an astonishing book published a few years ago called *The Entrepreneurial State*, Mariana Mazzucato, whom we have already met in connection with the history of the concept of value, took a closer look at the interactions between the market and the state in the case of important innovations. Taking Apple, one of the most valuable companies in the world, as an example, she shows that much of the technology on which the success of its most successful product, the iPhone, is based—such as the internet, GPS, touchscreen technology, high-capacity batteries, and the software for its voice-operated assistant, Siri—was the result of exploiting publicly funded basic research. Apple's legendary boss Steve Jobs may have been a marketing genius, and his people may have been geniuses in design, but in terms of technology, Apple simply assembled what already existed, because states had actively promoted their development. Mazzucato concludes that, quoting Steve Jobs, the 'foolish developer of innovations' is, in reality, the state.

Mazzucato writes, 'Most of the radical, revolutionary innovations that have fuelled the dynamics of capitalism—from railroads to the internet, to modern-day nanotechnology and pharmaceuticals—trace the most courageous, early, and capital-intensive 'entrepreneurial' investments back to the State.'[4]

Critics might argue that this was often driven by military interests. But even if that is the case, it does not change the reality of the state's role in the development of major technological innovations.

Companies such as Apple do not like to be reminded of the fact that their great economic success is built on social structures that are maintained by the state; nor do they like to hear that this, if nothing else, is a reason they should at least pay the taxes the state is entitled to claim from them.

The British Fair Tax Foundation estimates that the six biggest Silicon Valley companies—Apple, Amazon, Facebook, Google, Microsoft, and Netflix—cleverly manoeuvred to avoid around $100 billion in taxes between 2010 and 2019.[5] Amazon alone managed to gain a tax rebate from the American authorities of $129 million in 2018, a year when the company posted profits of more than $11 billion. For years, its tax rate was around 3 per cent.[6]

Companies such as Airbnb make use of publicly financed infrastructures without accepting any of the responsibility for maintaining them. For those who live in an attractive city—especially one that is well served by cheap airlines—it may at first seem like a good idea to rent out their apartment to tourists via this platform, or to rent or buy a property with the express intention of doing so. But when more and more people do the same, and the subsequent increase in rents means that hardly any local people can afford to live in their own neighbourhoods, the result is often the emergence of pretty-but-soulless districts that no longer have an authentic appeal for tourists.

That's the tricky thing about the tyranny of small decisions: it is, by definition, not guided by any superior authority that has the responsibility of taking a broader view and monitoring whether the sum of the individual interests really is resulting in the benefit of all. In other words, an authority that would prioritise the wellbeing of the group over individuals' ability to maximise their own personal benefit, which, in many cases, would end up securing the wellbeing of the advantaged themselves over the long term. This is called 'securing the public interest'; it requires a long-term perspective, and is one of the original tasks of the state.

Karen Vaughn, a professor and senior fellow of the Friedrich Hayek Program at George Mason University—not someone who could be suspected of being hostile to the market—wrote on this issue:

> [O]ne could easily imagine a spontaneous order in which people were led as if by an invisible hand to promote a perverse and unpleasant end.

> The desirability of the order that emerges as the unintended consequences of human action depends ultimately on the kind of rules and institutions within which human beings act, and the real alternatives they face.'[7]

On the role of the state, John Maynard Keynes concludes:

> The most important *Agenda* of the State relate not to those activities which private individuals are already fulfilling, but to those functions which fall outside the sphere of the individual, to those decisions which are made by *no one* if the State does not make them.[8]

Keynes also did not see state interventions in the market as exceptions to the rule, but rather as normal actions taken to preserve the balance between supply and demand—not only in the market for products and services, but also, for example, in the labour market, the ratio between exports and imports, the money supply, and the currency markets. And I would also add to that list instances when nature or future generations cannot defend themselves against overexploitation and discrimination.

The question is: Does the state even still know this?

And if so, does it have the courage to act on that knowledge?

Let's take a simple and clear example that we are all familiar with from our everyday lives: returning things we order online.

A group of researchers at the University of Bamberg calculated that people in Germany returned one-sixth of the parcels in which they received online purchases in 2018—because the goods were not what they hoped for, or they found them more cheaply elsewhere, or they were the wrong size, or because they wanted to experience the items in real life before deciding whether to keep them.[9] That accounted for 280 million parcels in one year. According to the estimates of the 139 retailers interviewed by the researchers, a small fee of less than three euros per parcel would have resulted in about 80 million fewer returned items in the year in question. The amount of fuel that this would have left unused would have reduced climate-harming carbon dioxide emissions by 40,000 tonnes, corresponding approximately to the total carbon dioxide emissions produced by 4,000 people in Germany in an entire year. Therefore, a return fee of less than three euros would have meant the equivalent of 4,000 Germans living in a completely climate-neutral way.

The companies that already charge a return fee are mostly small and medium-sized retailers. They registered almost no reduction in sales due to their imposition of the fee. None of them saw their profits fall, because they no longer had to bear such a large proportion of the costs of returning items. Most of the other small and medium-sized retailers interviewed said they would like to charge a return fee, but did not dare to do so, as they feared it would put them at a disadvantage over their competitors. Their plight demonstrated that without intervention by government, the market will never agree to the general imposition of a return fee. That could only be achieved by means of state regulation.

Big online retailers such as Amazon or Zalando, for instance, might not like that, because their size means they can shoulder the costs of returns more easily than small companies, and the current returns policy helps the big players make it more difficult for their smaller competitors to enter the market.

People who like to make a lot of their purchases online, and therefore also return a lot of items, might also not like the fact that such a fee means that they would have to go back to trying goods out and considering whether to really buy them before placing an order.

However, this doesn't change the fact that, on balance, introducing such a fee makes absolute sense. It protects the environment, is supported by the majority of retailers, and does not disadvantage anyone more than anyone else, since it would apply equally to all. The state would only need to make the decision to introduce it. Apart from the state—just as John Maynard Keynes argues—there is no other institution that can do so.

Do you know who else put this in a nutshell? It was Franklin D. Roosevelt, the US president who introduced the New Deal in 1933 to help the economy recover from a devastating economic crisis. In a 'fireside chat' with the nation, he said:

> The unfair 10 per cent could produce goods so cheaply that the fair 90 per cent would be compelled to meet the unfair conditions. Here is where government comes in. Government ought to have the right and will have the right, after surveying and planning for an industry to prevent, with the assistance of the overwhelming majority of that industry, unfair practice and to enforce this agreement by the authority of government.[10]

That's an interesting idea, isn't it—the state and market participants working as a team to put the development of a branch of industry on the right track by introducing clear rules?

Political freedom and individual responsibility also went hand in hand in classical economic theories of the state and market. So-called ordoliberal economists would say that decision-making and responsibility-taking belong together. The German constitution states that, 'Property entails obligations'.[11] That connection is increasingly becoming lost in our globalised, financialised, and digitised world. In his book *In schwindelerregender Gesellschaft* (*In Dizzying Company*), the business ethics professor Thomas Beschorner describes this failing as 'the balance disorder of the modern world'. He calls it 'halved liberalism', in which the state and the market no longer adequately fulfil their complementary roles. Beschorner believes that the function of political order in connection with the markets is not only economic in nature, but also ethical. Its task is to provide stimulus, rein in self-interested acts, and promote moral behaviour.[12]

The state and the market cannot be separated from each other. And there is no 'Mister Market' telling us what to do and demanding that we bend to his will. Well, I've never met him, at least. Have you?

Nonetheless, for years now, this 'halved liberalism' has passed on the responsibility for stopping the global destruction of the planet to individual citizens and their consumer decisions. If you want to do something for the environment, you have to consume sustainably. This is nothing other than the privatisation of environmental protection. That's good news for business, because it means companies can now offer environmentally responsible consumers an extra choice, with the right labels and the right credentials to soothe their conscience. And it's good news for politicians, because it means they can avoid the unpleasant job of introducing regulations, or even prohibitions, in the face of possible resistance.

How far have we come with that?

Despite the recent proliferation of organic supermarkets springing up everywhere in Germany, and the fact that even discount supermarkets now stock organic foods, the share of organically produced food on the German market is still well below 10 per cent. If we look at organically farmed meat, the situation is even more depressing. It accounts for 4 per cent of sales, at the very most, and, depending on the type of meat, is often far closer to just 1 per cent.[13]

Might this be because fewer than 10 per cent of the population can afford to buy organic produce, and even fewer can pay for organic meat, in one of the richest industrialised nations on Earth?

I don't think so.

The reason is that the market for agricultural produce is organised in such a way that it tends to reward unsustainable behaviour, and, indeed, even makes it more difficult for us to act sustainably. As we saw in the chapter on consumer behaviour, the price of many goods does not reflect the true cost of their production, and the same is unfortunately true of food products.

Can you guess what that means?

That's right.

Sustainably produced food is not too expensive, after all.

It's industrially produced food that is too cheap. And our consumption of meat is way too high. For the health of humans, animals, and the planet, it's simply far too high.[14]

And do you know what would help?

Reforming the system of agricultural subsidies. That would immediately reduce the price difference between industrially produced and organically produced food.

However, as is so often the case, it is worth looking at the issue from another perspective, since, without eating any less than before, we now spend far less on food than we used to. Over the past fifty years, the proportion of household income spent on food in Germany has fallen from 25 per cent to 14 per cent. On the other hand, since 1993, housing costs have risen for almost everybody except the richest 20 per cent—that percentile now spends 9 per cent less on accommodation than in 1993. At the other end of the scale, the poorest 20 per cent have seen their housing costs rise from 27 to 39 per cent of their income. That's connected to the question of whether people live in rented accommodation or own their homes. Rents have skyrocketed over the past ten years, but that's coupled with the fact that incomes at the lower end of the scale have fallen in real terms.[15] Bye-bye trickle-down effect! And welcome to a world where the relationship between the different amounts that people spend on essentials is suddenly turned on its head.

So does that mean organic food, which thrives in a sustainable agricultural system with more biodiversity and healthier soil, fixes more carbon dioxide, and has a lower impact on our groundwater, is too expensive? Will housing soon become a luxury? Or do we perhaps need new agricultural, minimum-wage, and housing policies to counter the explosion in land, rent, and consumer prices since 2010? All those trends could be influenced at the same time with a fiscal policy that is oriented towards the common good, but where is that policy? You remember the difference between creating and skimming-off value?

I would ask you to consider again carefully how monetary value and prices are put together. They are anything but value-neutral numbers. Every time a worldly phenomenon is converted into numerical terms, a value judgement is involved. And every value judgement influences what we pay attention to and what we consider when making decisions, and assessing political developments and how fair they are—politics always plays a part in the development of prices.

So the question is not whether incentives,

prohibitions, or price increases should be allowed or not. The question is which of those instruments have ceased to be effective in the new reality, or are being wrongly implemented, and are therefore standing in the way of our reaching the necessary goal of sustainable living. The market is not a rules-free space, but was in fact created by rules in the first place. Those rules have an influence on which liberties we enjoy and which we do not, on what is forbidden and what isn't, and on which innovations are probable and which aren't.

Were that not the case, the highly lucrative slave trade would probably still exist, and workers would have no right to demand an eight-hours-a-day, five-day working week.

In his book *Whose Freedom?*, the American linguist George Lakoff points out that freedom goes both ways: there is not only freedom from something, but also freedom to do something; and he then argues that it was state interventions that asserted those freedoms from the beginning of the Enlightenment onwards. State regulations led to the freedoms enjoyed by science and researchers, and they led to the expansion of universities, to public healthcare, to freedom of speech, opinion, and assembly, and to the equality of all citizens before the law. A financial market without state regulations and guarantees would also be unthinkable.

Otherwise, why would anyone give you access to a house to live in, in exchange for a pile of papers with words printed on them saying that a couple of digitised numbers will be transferred from your account every month?

It happens because the state not only sanctions breaches of contract, but also guarantees that those numbers represent a real entitlement.

That explains why the notes issued by the Bank of England still bear the sentence, 'I promise to pay the bearer on demand the sum of five/ten/twenty pounds.'

And let's be honest—on the roads, all drivers accept that their individual freedom necessarily ends at the point where the health and safety of others, as well as their own, is put in danger, just as all motorists also accept that state regulation is necessary to define the limits of drivers' freedom to behave as they want on the roads.

Why should it be any different on the road towards a sustainable economy?

In a now-famous article published in 1968, the US ecologist Garrett Hardin described a mechanism he called 'the tragedy of the commons'. In this context, 'commons' is taken to mean any open-access and unregulated resource. The example Garrett Hardin gave was that of common pasture land used by local cattle-herders to graze their cows. Since the commons had no owners, no one could be excluded from using it, and so everyone grazed as many cows on the commons as they wanted, for as long as they could. Everyone prioritised their own short-term gains over the long-term usability of the common resource. This overgrazing resulted in there not being enough grass left for everyone's cattle to graze on. Overexploitation by individuals to the detriment of all—that is the classic result when each person acts like a *Homo economicus* in an unregulated space. In view of this, it is actually surprising when markets do not fail: they generally only function well for classic barter transactions.

For common resources, at least, the overwhelming majority of economists now take the stance that the overfishing of our seas, the overfertilisation of our soils, and the illegal clearance of our rainforests make state intervention necessary to define the rules of use. The most contemporary and perhaps most important example of this is our use of the Earth's atmosphere as a carbon dioxide dump.

It is not possible to own a piece of the Earth's atmosphere, nor is it possible to exclude anyone from its use. The carbon dioxide that one person, company, or state releases into the air rebounds on everyone in the form of climate change. To attach a price to carbon dioxide emissions that would be sufficiently high to limit this unfair practice in the short term, and to end it in the medium term, is precisely the forward-looking responsibility the state is meant to shoulder. In this context, it is not only important to support companies that do act fairly. What we need is an entirely new deal that is not fixated on whether individual measures increase individual costs, but that instead takes a holistic view of the way we determine the cost/price of those things that form the main bases for a good life.

The market is simply not able to solve every problem when goods become scarce. And the actions of the state do not always result in fewer freedoms; indeed, it is often state action that makes such freedoms possible in the first place. If we are to solve the problems posed by the new reality, we must move away from 'halved' patterns of thinking. Global approaches must be found for scarcities of goods that span the planet, no matter how difficult that might seem.

9
Fairness

'We talk a lot here about giving more. We don't talk about taking less. We talk a lot here about what we should be doing more of. We don't talk about what we should be doing less of.'

Anand Giridharadas, journalist and author[1]

A couple of years ago, the ecologist Stefan Gössling decided to investigate the air travel habits of famous people.[2] He wanted to find out what impact such people have on climate change. Interestingly, no one had thought to do this before. In our society, celebrities represent those who have made it, as we often say. They are seen as people to look up to, as role models. They include artists, actors, athletes, business leaders, and politicians, for example, but also increasingly people who are not famous because of their job, but whose job it is to be famous. As so-called influencers, they take money from companies in exchange for giving their brands publicity.

Gössling analysed the air travel movements of ten such celebrities in 2017, from Microsoft founder Bill Gates and Facebook boss Mark Zuckerberg, to the singer Jennifer Lopez, the hotel-chain heiress Paris Hilton, the talk show host Oprah Winfrey, and the fashion designer Karl Lagerfeld. Although such people might be expected to be highly secretive about their movements, Gössling harvested the data he needed from their public social media profiles. Many famous people publicise their travels and the reasons for them, on platforms such as Twitter, Instagram, and Facebook. Some even see it as an essential part of curating their public image. They make a public display of a lifestyle that shows they are super-rich, which also implies the converse: that anyone who is super-rich must live like them—as if there were no other image or model of how to acquire money and how to spend it.

At the time the study was undertaken, the ten celebrities at its focus had a joint total of 170 million fans following their lives on the photo-sharing platform Instagram alone.

'Followers, and in particular younger people, may embrace frequent flier identities as a social norm established by celebrities,' the study says.

Bill Gates topped the celebrity frequent-flyer list in 2017, having spent at least 350 hours in the air. Since he primarily travelled by private jet, his flights emitted more than 1,600 tonnes of carbon dioxide into the atmosphere. Paris Hilton and Jennifer Lopez, second and third on the list respectively, also mainly travelled by private jet, and their trips emitted 1,200 tonnes and 1,000 tonnes of carbon dioxide respectively.

What does that have to do with fairness?

In the past, it was easy to believe that the lifestyles of the wealthier portion of the world were completely disconnected from the lives of the poor and extremely poor. One set of people were rich and the rest were poor, and if anything was to change, it was the poor who had to strive to become rich. What did the wealth of some take away from the lives of others, people thought.

However, that relationship can now be well expressed in numbers, since science has gained the ability not only to identify climate change precisely, but also to predict accurately the amount of emitted carbon dioxide likely to raise the mean global surface temperature by a given amount, and what effect that temperature rise is likely to have on our planet.

At a climate conference in Paris in 2015, almost every country in the international community agreed to limit global warming to 'well below 2 degrees Celsius' compared to pre-industrial levels. Scientific reports in the ensuing years showed that the extent of climate change and the cost of adapting to it would be far lower if global warming could be limited to 1.5 degrees. Calculations showed that this 1.5-degree limit meant humans could still release approximately 420 gigatonnes of carbon dioxide, calculated from the end of 2017. However, since we currently emit up to 42 gigatonnes a year, we now have less than five years until that allowance is used up.[3] After that, humanity will have to be practically climate-neutral if it is to keep to the agreed limit, meaning that any new emissions must be able to be cancelled out by the ability of nature and the oceans to absorb them. We have less than five years to accomplish probably the greatest economic, technological, and social transition in history.

This is, to say the least, a very, very tight schedule.

Using the figures at the beginning of 2020 as a basis, and averaged out over the world's population, this means that every person has to limit their emissions to no more than around 42 tonnes of carbon dioxide if the Earth is to warm by no more than 1.5 degrees.

Which brings us back to Bill Gates.

According to the Forbes list of the world's richest people, Bill Gates has a net worth of around $108 billion, making him the third-wealthiest person in the world.[4] In one year, he used up the equivalent of thirty-eight people's lifetime budget for carbon dioxide emissions—that means all the emissions for heating, mobility, and consumption they can still make while we remain under the 1.5-degree limit. One single person, using the allowance of thirty-eight. Just for himself. Just for the flights he publicised on social media. In just one year.

We currently still have a situation in which some people's lifestyle means they hardly make use of their theoretical carbon dioxide budget, and can make it available to others. Of course, different professions are also associated with different amounts of carbon dioxide emissions. Also, some families are scattered across the globe. So finding a formula for fairness is not easy here. But it never is.

What clearly is unfair is the fact that hardly any of those extreme emitters show any sign of questioning their lifestyle in any serious way. And the only possible rationale I can see for this is that they have the financial means to secure the dwindling remaining resources for themselves. Those are the same financial means that will then allow them to do something that all the people whose budgets they have used up with their private flights cannot do—adapt to climate change, move to places where it's still nice to live, pay the rising prices for scarcer food resources, and insure their homes against climate-based destruction. It is not until we add all their emissions over the past thirty or forty years, when the facts about climate change and its causes were already known, that the reckoning becomes really clear. These people's lifetime carbon dioxide balance is so off the scale that we would have to suck thousands of tonnes of carbon dioxide out of the atmosphere every year until 2050 to reduce it to the approximate equivalent of the emissions caused by the average world citizen.

Do you think that's fair?

So now you see the connection.

As I have shown in the earlier chapters of this book, we now live in the reality of a world that is full, in which we must adapt to the limitations of our planet. However, the fact that the Earth has limits has not yet entered human consciousness, and it does not yet determine our behaviour. Most of our attempts somehow to deal with those limits have not really helped us adjust our way of life to them.

I believe there is a perfectly simple reason for this. If we accept there are limits, we must also accept that goods and pollution rights are finite. If the cake cannot just keep on getting bigger, the question inevitably arises of how it is to be shared out. And if ecosystems can harbour only a certain amount of resources and can absorb only a certain amount of waste and fumes, the question must also inevitably arise of who can consume, discard, and emit, and how much is permissible. Environmental questions are always about distribution, and distribution is always about fairness.

I have already detailed some of the arguments that are made when the issue of fairness comes up in the public debate:

- Economic growth will create fairness;
- More efficient technologies will create fairness; and
- More sustainable consumption will create fairness.

I also detailed the fact that all these arguments turn out, on closer inspection, to support a narrative that says the cake *can* keep on getting bigger, and that the narrative turns out to be just a convenient story from a pseudo-reality that has no basis but which we continue to cling on to. Without directly attributing evil intentions to those who have told those stories and/or continue to tell them, we must simply say they are wrong. However, the fact remains that it is because of those stories that the questions of how the planet's resources can be more fairly distributed, in view of their limited nature, have not been answered, but have been left for future generations to deal with. And the fact also remains that it is predominantly the people who profit from the promulgation of those stories who are the same ones who have already profited disproportionately from the exploitation of the planet's resources.

Nonetheless, we are still constantly told that ecological aims run counter to social aims.

How often I have sat on discussion panels where a kind of relief seemed to set in when these 'deep-rooted conflicts of aims' cropped up, because it meant the debaters could continue to do nothing but contemplate how difficult everything is? And, obviously, that means immediate action is not possible. Really poor countries and people are rarely represented on such panels, although they are the ones who will bear the brunt of the social and humanitarian disasters in the case of an environmental catastrophe. This means that continuing to do nothing can be painted as socially considerate of the financially weak in the countries of overconsumption. I almost got the impression that politicians and corporate leaders, but also trade unionists, were incredibly grateful for the yellow jacket protests in France in early 2019, which were triggered by a rise in fuel tax as part of the government's move to a greener energy policy. It showed, they concluded, that real climate policies are something the population simply does not want.

But where was the question of what kind of climate policies the population *does* want? And of how they can be included in future-oriented social and progressive programs in such a way that the alleged conflict between environmental and social aims can be turned into a marriage of objectives? Where was the insight that social justice can be improved from the other direction, namely from above? How can we persuade people to support the enormous changes that introducing a sustainable economic system would bring with it, when we can't even guarantee that those changes will affect everyone equally? When the French government cuts wealth tax, that confidence is at least very strained.

In other words, how will we ever solve the problem of the environment if we don't begin to understand it as also being a social issue?

In the early nineteen-seventies, the American John Rawls, one of the most important political philosophers of the last century, took a new view of the issue of distribution. He believed the basic problem of the world was that the decision-makers, the rich and powerful, saw no advantage to any distribution of power and resources that differed from the current one. On the other hand, those who would benefit from a change in distribution, namely those with less wealth and power, had no or insufficient influence to bring about any change. The result was an ever-intensifying and essentially unsolvable dilemma of fairness.

Rawls illustrated this dilemma with a thought experiment involving people making decisions about their society from behind what he called a 'veil of ignorance'. Behind the veil, people can still think rationally, but—a little like before they are born—know nothing about what situation they will be end up in. They don't know what skin colour they will have, what gender, what nationality, what family background, and—this is an additional unknown I would like to add—what generation they will be born into. They might come into the world as Bill Gates' child or as the child of a rice farmer in Bangladesh. However, in this thought experiment, it is unlikely that an individual will end up as the child of one of the richest humans on the planet; becoming the daughter or son of one of the poorest people on Earth is far more likely. That is quite simply because there are so many more poor people in the world than rich people.

The question that Rawls asks in connection with this thought experiment is, 'What choices would people make about the kind of society they want to live in, if they did not know before making those choices what position they would occupy in that society?'

This is basically taking what is called a systemic perspective. Doing so enables us to formulate progressive goals and come up with multiple approaches, rather than considering each individual measure in isolation. Rawls believes that very one of us has an intuitively correct sense of what is fair and what is unfair. And science now even has figures to support that view.

In 2012, the psychologist and behavioural economist Dan Ariely, together with his colleague Mike Norton, asked Americans how they believed wealth should be distributed across society, and how they believed it was in fact distributed.[5] The researchers asked respondents to imagine society divided into five groups with an equal number of citizens in each group, ordered according to wealth. Those questioned were then asked to predict what percentage of overall wealth is available to the people in each of the five groups. On average, the respondents had the following view of an ideal distribution of wealth in America: the richest fifth would have just over 30 per cent of the wealth at its disposal, and the poorest quintile would have 11 per cent. There was no significant difference in the answers given by respondents of different genders and of different political persuasions—that is, between Democrat and Republican voters.

When the respondents were asked to estimate the actual distribution of wealth in their society, they thought that the richest quintile held almost 60 per cent of the wealth, while they thought the poorest fifth held less than 5 per cent. In fact, at the time the research was carried out, the richest fifth of American society held almost 85 per cent of the wealth, while the poorest fifth owned less than 1 per cent.

Or, to put this another way: American society is far less equitable than it appears to those who live in it.

And since the survey was done, inequality in the distribution of wealth has grown so much that more recent studies now give figures for the top 1 per cent separately. That percentile now holds 40 per cent of household assets in the US.[6]

And the situation is virtually the same the world over.

The World Inequality Lab studies the dynamics of global inequality. Its 2018 World Inequality Report, drawn up in collaboration with more than 100 researchers from around the globe, shows that the inequality between rich and poor has increased around the world since 1980.[7] In that period, the richest 1 per cent of the world's population was able to secure more than a quarter of the global increase in wealth for itself. And in those almost-forty years, the richest 0.1 per cent has increased its wealth by about the same amount as the bottom 50 per cent.

Whichever way you look at it, the fact remains that of the wealth generated by the economic growth achieved since the beginning of modern globalisation, a small amount has reached a large number of poor people, an unbelievably large amount has ended up in the hands of a very few rich people, and little or nothing at all has gone to the large middle class.

The World Inequality Report also shows that the gap between rich and poor is less extreme in countries that pursue active distribution and social policies. This means that faster progress in eradicating poverty is made when it is a stated political aim and not just a possible side-effect that might occur while we all wait for the uncontrolled trickle-down effect to kick in and, allegedly, raise all boats.[8] What if we were to halt the process of constant growth to reduce inequality, and started managing the distribution of goods, resources, and opportunities in such a way that it begins to resemble people's ideals?

A first step could be, for example, a one-time allocation of 10 per cent of global Gross Domestic Product to the development of healthcare systems, educational institutions, resilient agriculture, and renewable-energy provision for people with little purchasing power of their own.

That would be $8.2 trillion.

Far too much?

Where is it supposed to come from?

The economist Gabriel Zucman estimates precisely that this sum is the amount of money currently hidden by the world's rich in tax havens.[9] Tax increases are generally imposed so that the increased government revenue can be invested for the common good, aren't they? Assuming a one-off tax of around 30 per cent—a level quite normal in many countries—were levied on those funds, $2.7 trillion would end up in the global public purse. As Zucman makes it clear in his book *The Hidden Wealth of Nations*, with a budget of that size, the international community could make huge investments in essential public services.

How can we redress these huge imbalances? Or, as a first question: Why aren't we more honest about these huge imbalances when we talk about how to solve them?

Let's look again at someone like Bill Gates, for example. People like him are a good illustration of the difficulties that arise when we try to treat the symptoms of a faulty system rather than looking at the system itself. Gates did not inherit his fortune. His wealth is the result of his entrepreneurial ingenuity—almost everyone in the world works with or has at least heard of Microsoft products. The private foundation of Bill Gates and his ex-wife, Melinda—the biggest private foundation in the world, measured by its endowment of $53 billion—has been funding the development of vaccines against diseases such as AIDS, tuberculosis, and malaria for many years now, as well as trying to improve agricultural development in Africa. Indeed, they spend more on healthcare, education, and food-supply projects than many a democratically elected government does.

Surely this means that the carbon dioxide emissions that Bill Gates creates with his private jet are well invested? Isn't it wonderful that he's taking on the problems that the national governments of this world ignore?

Or course, it is marvellous that someone is tackling these problems. But, while national governments must answer to their opposition, the law courts, and the electorate, Bill Gates's foundation decides for itself who and what to support and collaborate with. It defines its own procedures and conditions of cooperation. And it makes its own decisions, according to the organisation Global Justice Now, to smooth the way for chemical giants such as Monsanto and global grain-traders Cargill to enter African markets, or to invest in companies such as Monsanto and McDonald's by buying their shares.[10]

In his 2018 book *Winners Take All*, the American journalist Anand Giridharadas examined how this kind of philanthropy has established itself as a modern version of the medieval practice of selling indulgences, in which spiritual favours were distributed by the Church in return for money. It does not aim to bring about any real change in the political situation, the distribution of wealth, or the privileges enjoyed by those who take part in it.

'The winners of our age don't enjoy the idea that some of them might actually have to lose, to sacrifice, for justice to be done,' writes Giridharadas. 'You don't hear a lot of ideas involving the privileged and powerful actually being in the wrong, and needing to surrender their status and position for the sake of justice.' They don't mind being challenged to 'do more good', and thrive on the gratitude they receive when they do, Anand Giridharadas continues, 'But never, ever tell them to do less harm.'[11]

Generosity is not justice.

Redistribution always sounds as if it means some people have to give up something they are entitled to, while others — supposedly less successful, less intelligent, and less efficient — should rely on a kind of patronage. But it is highly unlikely that, since 1980, top company executives have become on average 1,000 per cent more intelligent, efficient, and hard-working while their company employees have only improved by an average of 12 per cent. That's precisely how the figures for income distribution have developed in American companies since 1978.[12] And after all his empirical legwork on capital in the twenty-first century, Thomas Piketty saw the reason for the growth of inequality less as a result of the exploding productivity of company executives than of state taxation policies. He points to the fact that the highest earners of one company are board members of other companies, and so the two boards end up policing each other's pay structures.

Fairness is not just about the fair distribution of wealth; it is also about equal opportunities. That includes the equal opportunity to lead a dignified human life, as well as an equal opportunity to influence the conditions of that life.

The same principle applies to nation-states.

* * *

Some time ago, the World Resources Institute published a diagram breaking down by country carbon dioxide emissions since the beginning of industrialisation.[13] It shows that, as the biggest emitter, the United States was responsible for a total of 27 per cent of cumulative global emissions between 1850 and 2011, followed by the countries of the European Union, including the United Kingdom, which produced 25 per cent of emissions. Countries such as China, Russia, and India, for example, lag a long way behind. Of course, this puts a different perspective on the fact that whatever we in the global North can, or at least could, do about global warming is cancelled out by the huge appetite for energy in our countries.

As that chart shows, for a long time now, we have been financing our leap in development with a loan from the global climate, and humanity will be paying that mortgage back for a long time to come. If we are not to conclude from this that equality can only be achieved when other countries emit at least as much carbon dioxide as the US—and, in the meantime, the US is not going to suddenly stop living as it has—then the only alternative is to find some kind of a trade-off between different nations.

What form might that trade-off take?

Take the Amazon rainforest, for example. According to figures published by the Helmholtz Centre for Environmental Research, up to 76 billion tonnes of carbon are stored in the Amazon forests, and every year they bind an additional 600 million tonnes.[14] The rainforest in an important factor in combatting climate change, and is therefore also important for the entire international community. As such, it was not surprising when the French president Emmanuel Macron expressed concern at the thousands of forest fires destroying large parts of the planet's green lung in 2019.[15] On the other hand, most of the Amazonian rainforest is in Brazilian territory, and former resident Jair Bolsonaro considered any suggestion from foreign heads of government that Brazil should act more quickly and decisively to put out such fires as interference in the internal affairs of his country.

A classic conflict.

Brazil is anxious to catch up economically with the GDP-celebrating Western industrialised nations. Countries with Brazil's per-capita income level are called 'emerging economies', as they are on the verge of joining the club of countries that have 'made it'. To achieve this aim, Brazil is willing to use the resources of the rainforest—its wood, the mineral deposits thought to be beneath its soil, and not least the agricultural opportunities offered by the land the forests occupy, which is often turned first into pasture land for cattle and then into crop land for soya. Brazil is the world's largest exporter of beef[16] and the second-largest exporter of soya,[17] which, as mentioned previously, often finds its way to Germany in the form of animal feed to fatten up cows and pigs. Europe is in the process of negotiating further trade deals with South American countries in the so-called Mercosur trading bloc, to make such exports and imports even easier.

So, by moralistically wagging its finger and issuing threats, does Europe make itself look strong, or, more likely, ridiculous?

And who can really blame Brazil for its plans?

Countries such as Germany and Britain exploited the resources they discovered in their own territories as they saw fit, with no interference from abroad. If they had left their coal deposits in the ground, for example, we would be dealing with far lower amounts of carbon dioxide in the atmosphere today.

In his 2002 book, *Kicking Away the Ladder*, the South Korean economist Ha-Joon Chang described how the industrialised nations of the global North want to ban developing nations from using the very methods they used to fuel their own economic rise: high tariffs to protect their domestic economies; product piracy; and concentration on key industries are all examples of methods that were used, and are being used again, by countries such as the US, Britain, Germany, and Japan to achieve more economic growth.

'It is a very common clever device that when anyone has attained the summit of greatness,' writes Chang, 'he kicks away the ladder by which he has climbed up.'[18]

How can we get out of this race to destroy the world? How can we come to an understanding of fairness that allows us to start working with each other again, rather than against each other, and to reconcile our social and environmental aims?

For me, the key to this is thinking from a future perspective. And thinking systemically. As we learned from John Rawls' veil of ignorance, we all have pretty similar individual concepts of what constitutes fair distribution, once we step back from directly comparing ourselves with one another.

If we extrapolate the idea of a fair distribution of wealth expressed by the Americans in Dan Ariely's survey, we see that the 10 per cent of GDP which those respondents said should be in the hands of the poorest 20 per cent of the population would be the same amount globally as the $8.2 trillion currently stashed away in tax havens. That works out at a little over $10,000 per person per year. Or $27 a day.

So the idea of setting the target of $7.40, or even $15 a day, as the measure for extreme poverty, as I discussed in chapter four, doesn't seem that far-fetched after all. The World Bank's baseline of $1.90 per day certainly does seem off the mark. A system that can designate someone with an income of over $1.90 a day as no longer living in extreme poverty, without considering the picture at the other end of the scale, can only have been created, as I suspect Rawls would see it, by those at the top of the pile.

The slogan of the UN's Sustainable Development Agenda is 'Leave no one behind.' In view of the fact that our planet's resources are limited, the reverse is also true: 'Let no one run off ahead.'

This formula even brings us closer to the conditions under which an 'invisible hand' allows positive results to be negotiated between many parties on a permanent basis. As Oliver Richters and Andreas Siemoneit write in their book *Marktwirtschaft reparieren* (*Repairing the Market Economy*):

> Ownership guarantees responsibility and discourages neglect. Those functions are legitimate and important. But ownership cannot be seen as an absolute, because its primary function is social in nature, not individual: it aims to make the division of labour among strangers, rather than accumulation, possible. The limits of ownership must be drawn at the point where it restricts the freedom of others, that is, where it leads to the excessive accumulation

of power and inevitably enables some to reap who have not sown.[19]

In this regard, the idea that a few upper limits on wealth accumulation might also help the upper crust is not so absurd. The majority of researchers who study societies have compiled data in recent years confirming Tim Kasser's findings about the psychological effects of materialism—when society moves away from equal opportunities and towards expressing value only in terms of money, property, and fame. When that approach is adopted in a society where the distributions of opportunity and wealth are highly skewed anyway, even the most moneyed of social circles will suffer a great deal of stress.

In 2019, Daniel Markovitz—just like Michael Sandel—wrote a book on how this kind of meritocracy is damaging for everyone, and what top earners do to achieve and maintain their top earnings, top lifestyles, and conspicuous consumption. They devote no less than their entire lives and potentially their health to it. That dedication begins at kindergarten age, when their entry into elite institutions begins. A little later in life, the students at top-performing schools have stress levels three times higher than those of their peers at normal schools. A study at one high school in Silicon Valley found 54 per cent of students displayed symptoms of depression, and 80 per cent showed moderate to severe signs of anxiety.[20] Bankers are still expected to work from nine to five, but now it means beginning at 9.00 a.m. on one day and working through to 5.00 a.m. on the next. They lose sight of the meaning of their jobs, and miss out on time with their families and friends, and on taking care of their health. But if they are going to 'keep up with the top people', they must keep earning top money.

Do you know what 'upper limits' means in real terms?

Sufficiently progressive taxation and sensible competition legislation.

Under such conditions, fairness and equality become not just a social end in themselves, but also a means to secure a subjectively good quality of life and to promote social cohesion.

And how does the future-oriented perspective that emerges in healthy systems apply to environmental problems?

Taking the example of the Amazon rainforest again, one possibility is the kind of arrangement for Ecuador that reached a relatively advanced stage of negotiations under its-then president, Rafael Correa. It is based on the idea of a fund that rich countries would pay into, in exchange for which Ecuador would leave the oil reserves found in its Yasuní National Park untapped. Eventually, the scheme failed due to lack of trust: there were doubts about whether Ecuador would still forego the oil revenue after the money in the fund was used up. But that is a question of political will and reliable institutions—for example, ones that guarantee permanent financial transfers, secured by new technologies such as blockchains. A scheme such as Ecuador's does not represent a conflict of interests between social and environment goals. Quite the opposite, in fact.

Another idea in this regard is that of creating an Earth Atmospheric Trust, as suggested by the Nobel Prize–winning environmental economist Elinor Ostrom and her team.[21] Under this scheme, individuals who exceed their share of the carbon dioxide emissions budget would pay a corresponding amount into a trust fund. Part of the money would be returned to all people on Earth in the form of an annual per-capita payment. The remainder of the money would be invested in transforming our energy systems and other climate-protection projects. Poorer people would benefit from this scheme, since they tend to emit less carbon dioxide. There has been a similar suggestion for setting a price for carbon dioxide emissions in Germany, and its rejection was met with great incredulity by economists from all schools of thought.[22]

The European Union's mechanism called European Effort Sharing, as part of its emissions trading system, lifts directly from this model—and means that Germany has to pay penalties of up to 60 billion euros to neighbouring states if it fails to adapt its climate-protection policies quickly.[23]

These examples may seem rather unconventional to you now, but they are the kind of mechanisms that we will be in great need of in the future — I'm sure of that. For me, they are future-oriented because they include the idea that those who are wealthy today — and therefore have the power to do more — because of their intensive exploitation of resources in the past, must make use of that empowerment and act. It is their responsibility to do so, since the source of their wealth — development through the massive extraction of resources — is no longer available to others. That is justice rather than generosity.

We live in a time of crises, and in such times it makes a lot of sense to stop fixating on what we stand to lose as individuals. It's better to focus on what we can attain by sharing the resources we have. Consider, for example, the response to the flooding of the River Elbe in eastern Germany a few years ago: the crisis was identified, as were the options for action, and everyone contributed whatever they could: sandbags, tractors, trucks, shelter, their bodily strength, information, money, cups of coffee or tea, and sandwiches — just whatever each person was able to give.

And what do we do now?

We squabble over whether that sandbag is too big, this tractor too small, the truck too green, the neighbourhood a disgrace, the information not 120 per cent secure, the compensation payments too small for the distance of your house from the river, and the coffee too weak. Unfortunately, nobody ends up benefitting much—and the floodwaters continue to rise regardless.

But there is a better way.

Fairness is the key to a sustainable economic system that functions globally. It is the only way to prevent environmental and social issues from being played off against each other. The two belong together and can only be solved in concert with one another. This new kind of fairness requires us to slaughter a few of the sacred cows of current narratives of economic growth and to forge new paths. And that will also allow us to escape the increasingly rampant side effects of those narratives.

10
Thought and Action

'Although some have argued that today's age is one where "the great dream is to trade up from money to meaning", there is an unshakable and discomfiting sense that, in our obsession with optimizing our creative routines and maximizing our productivity, we have forgotten how to be truly present in the gladdening mystery of life.'

Maria Popova, author[1]

A few years ago, the Wuppertal Institute, where I worked at the time, held a series of seminars on making energy systems more sustainable. The invited participants were young decision-makers from around Europe who were facing the task of helping with the strategic organisation of that transformation in their companies, municipalities, governments, and civic organisations. We wanted to direct their attention to the quite simple fact that every time we make a decision, we do so under certain conditions. Those conditions form the framework of what we consider realistic, feasible, and desirable. They are like a box that contains us as we think and act in our day-to-day lives. During innovation processes, it is helpful to think outside that box briefly, to seek new ideas for new, small changes. But it also makes sense sometimes to step far enough out of the box so that we can get a view of the box itself. Perhaps the box needs changing?

The seminar's aim was to present the participants with the scientific evidence to back this up and to show them a new way of looking at things. We wanted them to develop a better understanding of how to use intelligence to initiate major transformation processes and how to have the patience to allow them to stabilise. I had the impression that by the end of each event, most participants couldn't wait to get back to their places of work and start implementing those new ideas.

That was the point at which we told them about the 'Mean Monday' phenomenon.

'Mean Monday' was the phrase we used to describe something that is familiar to anyone returning from such an event or presentation with fresh zest and inspiration, and a head full of ideas about how to change things and do everything differently from now on. And then they find themselves back at the same organisation, with the same targets, procedures, conversations, and meetings, and everything is the same as before. We wanted to make sure our seminar participants were prepared for that letdown.

And I want to make sure you are prepared for it, too.

If this book has had the effect on you that I hoped in writing it, you will have stepped out of the box for a moment. Perhaps you now look at the world around you with different eyes. It might be that you no longer believe certain narratives, now recognise different connections, and feel the impulse to reject some points that previously seemed to be accepted by everyone, including you. If this book has had the effect I hoped for, you may even have had some ideas of your own about we can move towards a more sustainable future. A future in which humans and nature are reconciled. A future in which there is relief from all those large and small driving forces that push us towards a way of life which today neither provides the greatest benefit for the greatest number, nor allows the basic means of life, on which everyone's happiness depends, to regenerate. A future in which we are better at sharing and have finally learned to be satisfied with what we have.

And then you look up, and the world is still the same. The people you interact with in various contexts are also the same, and they may even actively prefer to leave everything as it is.

That is your Mean Monday.

* * *

So how are we going to go about changing a rethought world?

After everything you have read in this book, I hope you will agree with me that change is necessary. Business as usual is not an option, because it will lead to radical and unwelcome consequences. Even if *we* don't change a thing, things will change drastically—and not for the better. Our economic system will not stand still while we spend the next thirty years arguing and eventually reaching agreement on what minimal changes we want to allow—as long as they don't interfere with our beloved principle of blind economic growth. We are all part of a network of systems in which everything has an effect, whether we like it or not, and whether we do things differently or not. On the flip side, that also means we have the chance to steer those changes in the direction we choose. More precisely, we not only have a chance, but also a responsibility to grasp that chance. Every day, every one of us can be part of the change we want for the world, even if the changes we make feel small and insignificant at first.

Of course, the world is not going to change just because you read a book. Nor, of course, will it change just because I wrote one. I'm neither the head of the government of a major industrialised nation, nor the president of a multinational corporation—and, I'm guessing, neither are you. But even if you or I were in such positions, that amount of power and persuasion alone would not be enough to put our systems back on a healthy footing. I often hear that even those with power and influence are trying to come up with binding regulations that can both guide innovation and investment in the right direction, and inspire confidence in people that they will continue to be effective in the long term. That is what I have called the social contract in this book. As citizens, we must demand this kind of acceptance of responsibility for the future common good—and we must accept it ourselves.

After all, democracy doesn't mean that all you can do is wait for election day, or that you have to be a government or a corporation to make a change in the right direction. Experience has also shown us that just because someone is the head of a government or a company, they can't be relied on to make the required changes. Changing things will take a large number of people who really want to do so, and that means everybody counts. It begins with a thorough reappraisal of the aforementioned box and a deep consideration of what course of action is sensible and expedient, and of what our criteria will be when choosing what to do next from all the possible doctrines, routines, and models.

In chapter two, in which I described how an astronaut took a photo of our planet from space, I showed how important our human view of something can be. It influences our approach to that thing, how we deal with it, and how we relate to it. Our images of the Earth, nature, our human—or inhuman—possibilities, what the aim of progress is, how technology should be utilised and what is fair and just, inform the way we infer what is possible and what is not.

My invitation to you was to question those images.

For me, rethinking our world brings a kind of liberation. Even if we cannot bring the conveyor belt of endless extraction to an immediate halt—that would be a very abrupt and rather undesirable change—we can develop the necessary clarity, creativity, courage, and confidence to turn it into a regenerative cycle. This doesn't have to be done by a powerful few; we can all do our part. That's why I use the word 'we' so often in this book.

Even if you disagree with me on many points, the networked systems we are all involved in mean we are all connected, even if we have different opinions. We can have horrible screaming matches about our differences, get angry and hurl a lot of insults at each other. We see that trend developing right now. Alternatively, we can decide to try to learn from each other, and to be honest about what is really important to us, what we are prepared to share, and what exactly we mean by certain concepts. It might seem silly to you, but that's more or less what our children learn at kindergarten. It's not a case of avoiding arguments at all costs. It's about reducing the amount of 'I' in our 'we'.

The louder someone claims that there is no alternative, the more closely you should want to question that claim. And, when I say closely, I mean closely. Don't accept being brushed off with rapid-fire answers, confusing statistical constructs, complex acronyms and abbreviations, or gobbledygook. And don't accept evasive answers such as, 'Those are the words of someone with a warm heart!' Your questioning is like a feedback loop, and can cause consternation, give food for thought, and have an impact. The way a society develops depends on the feedback loops at work in it. Even if you don't get a satisfactory answer straight away, you will have had a lasting effect. And we should all realise that we can make conscious use of such effects.

In times of crisis, questioning the box that shapes our thoughts and actions brings clarity about what aspects of the box could be different. And the more the box begins to wobble, the more courage is gained from the process of rethinking. And courage is something we all need a lot of these days.

In a democratic society, we rely not only on political actors to have the courage to make difficult decisions; we also depend on the population to have the courage to support them. It is going to be hugely difficult to break with the old habit of thinking of everything in monetary terms and retaining at the top position in our list of values what is only a means to an end. If there is something that gets far too little attention in the debate about the deep drivers of unsustainable development, it is the issue of the financialisation of our world and our relationships. There are so many other ways of organising and expressing the division of labour, cooperation, and the creation and assessment of value. Once again, monoculture only leads to extreme fragility. The central tasks of a sustainable society are no longer considered to be the responsibility of environment, development, and social affairs ministries, but of finance and economics ministries. They have sovereignty over all the numbers and concepts that make up the walls of the box.

And then there is the courage needed from consumers to use their purchasing decisions to support the present and continued existence of companies that are already developing tomorrow's innovations today. And the courage of the media to report in a more nuanced way on the aims and effects of legislation and the different faces of economic success. And the courage of corporations to include social and environmental added value in their balance sheets, and the courage of investors to prioritise such added value. Not forgetting the courage of mayors to plan their cities together with their citizens, and finally the courage of education ministers and school principals to include materials in their lesson plans and textbooks that teach the clarity, competence, and courage we need in the twenty-first century.

An experience of self-efficacy is the best stimulus for a person to switch from reactive defence to proactive solution-finding in a crisis. And if we employ our self-efficacy to create more mutual understanding and cooperation, more effective people will be mobilised to take action more quickly than you perhaps currently dare to dream of.

When we said goodbye to the participants in our seminars, we always gave them three pieces of advice before they returned to their old working environment in order to change it.

Always be kind and patient, but persistent. If you hit a dead end, take a step back and look at the box to see if there is another possible approach. There are many entry points for initiating change, and they may be found in ideals, language, statistics, incentives, processes, office design, or the culture of interpersonal interactions. Talks given by outside experts, examples of successful pioneers, and new alliances are also popular tools.

Seek allies. I promise you, once you out yourself as a rethinker you will find there are far more around than you thought. Find a language and way of interacting, or a completely new form or organisation, to express your goals. The more we break free of old concepts in our everyday lives, the more clearly the new way ahead will appear. There will always be several possible paths; when rebuilding the box, there is space for many heroes. Appreciating the different abilities and contributions of others is just as important as sharing and spreading the positive narrative of success, and opening up our arms when things go wrong.

Don't let the Mean Monday phenomenon get you down. There are more days in the week than Monday. It's important to be kind to yourself and to take the findings of psychological and happiness research to heart: your own inner drive for a cause is a more reliable motor than acknowledgement and approval from others. When awareness of a problem is not widespread or is not sufficiently great, the approval of others can often be minimal at first, especially in the case of processes aimed at bringing about profound change, such as rebuilding boxes. Personally, I'm very familiar with that feeling.

Concentrate on things that are within your power to influence, and don't worry too much about things that aren't—including bad reactions to your activities from those around you. Stay connected to your original intention. You cannot take on responsibility for any more than this, but that's already a lot. And, to end with a very important point: hold on to your sense of humour and ability to laugh about things; they should never die. Creating the future sustains us all.

Of course, no one can predict precisely what kind of a world will emerge when we let go of our outdated concepts and narratives. But if we combine that healthy dose of self-interest when making decisions with concern for the health of all, the tyranny of small measures will quickly be replaced by a different narrative: that the whole is greater than the sum of its parts.

Suggestions for further reading and action

The following is a small selection of suggested ways to get more involved or to find out more. I have tried to include suggestions in a variety of areas, to suit anyone's situation and interests. Most of my suggestions are for platforms rather than individual initiatives so as to provide as much choice as possible.

Further thinking

Tim Jackson, *Prosperity without Growth*, expanded and revised edition, Routledge, 2017.
A classic work that helped definite the debate. Jackson continues to develop his ideas along with colleagues at the Centre for the Understanding of Sustainable Prosperity (CUSP) at Sussex University.
https://www.cusp.ac.uk.

Kate Raworth, *Doughnut Economics: seven ways to think like a 21st-century economist*, Random House, 2017.
First publicised in 2012, the doughnut diagram showing the ecological ceiling and social foundation was a break-through in the debate around green economies at the United Nations level. In this book, Raworth presents a new kind of economy for the space between those two boundaries.
www.doughnuteconomics.org

Pavan Sukhdev, *Corporation 2020: transforming business for tomorrow's world*, Washington, DC, Island Press, 2012.
This German former banker and current WWF president led UN's The Economics of Ecosystems and Biodiversity (TEEB) study, before focussing his work on reforming consumption structures for more sustainable business practices.

John Fullerton, *Finance for a Regenerative World*, Capital Institute 2019–2021.
This former investment banker has founded a think tank to transfer the regenerative principles of biological systems to the design of economic solutions, including a re-invention of the financial system.
https://capitalinstitute.org/regenerative-capitalism/

Maja Göpel, *The Great Mindshift: how sustainability transformations and a new economic paradigm go hand in hand*, Springer, 2016.
An academic work on the connection between transformation research and concepts for a sustainable economy, as well as a template for the training methods of the System Innovation Lab 2016.
www.greatmindshift.org

The handbook from the System Innovation Lab
https://epub.wupperinst.org/frontdoor/index/index/docId/6538

The Wellbeing Economy Alliance
A global network of organisations and individuals engaged in research, experimentation, publishing, organising and increasingly networking on economies in the service of nature and humankind. This network also includes the first countries and regions to be serious about revising the way we measure prosperity.
www.wellbeingeconomy.org

Forum for a New Economy
Platform for a new economic paradigm
https://newforum.org

Evonomics (online journal)
The Next Evolution of Economics
www.evonomics.com

Further action

Consumption and everyday behaviour

Sustainable products

Utopia—product advice and background articles on living more sustainably: https://www.utopia.org/

Biome—Online shopping with sustainable brands: https://www.biome.com.au/

Greenpeace—Information about agricultural practice, policies, and politics: https://www.greenpeace.org/eu-unit; https//www.wri.org/food

Money as a resource

Fair Finance Guide—Information about banking practices: https://fairfinanceguide.org/

Forum Nachhaltige Geldanlage (Forum for Sustainable Investment)—Information about investing: www.forum-ng.org

Positive Money—A not-for-profit research and campaigning organisation based in London that works with economists, academics, journalists, and policy-makers to bring about a fairer money and banking system: https://positivemoney.org

Sustainable travel

https://sustainabletravel.org
https://www.sustainability.booking.com/

Atmosfair provides CO_2 offsetting: https://www.atmosfair.de/en/

Companies and organisations

A better economic balance

Economy for the common good: https://www.ecogood.org/

Benefit corporations
https://bcorporation.eu/

Global compact—a Sustainable Development Goals compass
https://sdgcompass.org/

New organisational forms

Regionalwert AG—Citizens' share companies connect investors with sustainable regional businesses across Germany: www.regionalwert-treuhand.de

Purpose Foundation—employee-owned stewardship:
https://purpose-economy.org/en/
https://entrepreneurs4future.de/

Political accountability

Stiftung 2 Grad (German CEO Alliance for Climate and Economy)—companies demanding political regulations for climate protection:
https://react-initiative.de—English button

Global Alliance for Banking on Values—Banks raising awareness of necessary regulations:
http://www.bankingonvalues.org/

Knowledge multipliers

Education

Global Goals Curriculum—to teach the skills necessary to achieve the UN Sustainable Development Goals, in collaboration with the OECD Learning Compass 2030:
https://www.ggc2030.org/en/home
https://www.oecd.org/education/2030-project/

Media

Selected content to guide you through the news jungle:
https://worldsbestnews.org
https://www.solutionsjournalism.org/

Enorm—Magazine for social responsibility
https://enorm-magazin.de/

Neue Narrative—Magazine for new ways of working
www.neuenarrative.de

Further restructuring

Comprehensive circular economy

Ellen MacArthur Foundation
www.ellenmacarthurfoundation.org

Cradle to Cradle
https://epea.com/en/about-us/cradle-to-cradle

Political change at the local level

German Zero—organising legislation for a 1.5-degree climate target and influencing local climate policy decisions:
https://www.germanzero.de/english

Ecovillages—worldwide network for local community strategies for regenerative development:
https://ecovillage.org/

Community-supported agriculture:
URGENCI is the international grassroot network of all forms of regional and local solidarity-based partnerships for agroecology (LSPAs), of which community-supported agriculture (CSA) is the best-known iteration. URGENCI is an acronym standing for An Urban-Rural networks: GEnerating New forms of exchanges between CItizens. As a social movement, Réseau International URGENCI brings together citizens, small-scale food producers,

consumers, activists, and researchers representing local solidarity-based partnerships for agroecology networks and initiatives in over 40 countries.
https://urgencinpet

Transition Towns—international network
www.transitionnetwork.org

C40 Cities—Network for climate protection connecting major cities around the entire world:
https://www.c40.org/

Political change at the national and European level

Deutscher Naturschutz Ring (German Nature Protection Circle)—Map of socio-ecological transformations:
https://www.dnr.de/themen/sozial-oekologische-transformation?L=46

German government strategy for sustainability:
https://www.dieglorreichen17.de/g17-de

Government dialogue Gut Leben in Deutschland (Living Well in Germany):
www.gutlebenindeutschland.de

SDG Watch—Civil society monitoring of the realisation of sustainability goals in Europe
www.sdgwatcheurope.org

European Progressives Sustainable Equality 2019–2024 Report:
https://www.socialistsanddemocrats.eu/publications/report-independent-commission-sustainable-equality-icse-2019-2024

Club of Rome—Planetary Emergency Plan
https://www.clubofrome.org/publication/the-planetary-emergency-plan/

WWWforEurope—Welfare, Wealth and Work for Europe project with ideas for new competitiveness and transformation:
https://www.wifo.ac.at/forschung/forschungsprojekte/wwwforeurope

Designing social innovations

Innocracy Conference of the Progressives Zentrum:
https://www.progressives-zentrum.org/innocracy2019/

Nesta Foundation in London:
www.nesta.org.uk

Acknowledgements

Writing a book as personal as this one is to me also requires very personal support, and I received that, as well as wonderful encouragement, from Uwe Schneidewind and Thomas Hölzl. Maria Barankow and Julia Kositzki at Ullstein Verlag were both magnificent and tireless in their support. My thanks go to them. I would also like to thank my team at the offices of the German Advisory Council on Global Change for all their patience in view of my irregular working hours. And I extend equally warm thanks to Jonathan Barth for his reflections on the chapter on growth. Tanja Ruzicska was not only an excellent editor with a clear view and a warm laugh, but she also gave me her unwavering support at all times. I awarded her the prize for positive psychology!

But the biggest thanks goes to my mother once again, who never hesitates to spring into action whenever too many parallel jobs threaten to overwhelm the system here. My father also gave well-meaning and generous support in 2019. You guys are amazing!

Notes

Chapter One: An Invitation

1 World Commission on Environment and Development, *Our Common Future* (the 'Brundtland Report'), Oxford University Press, 1987.

Chapter Two: A New Reality

1 Rachel Carson in her acceptance speech for the National Book Award 1952.

2 See: Apollo Flight Journal, https://history.nasa.gov/afj/ap08fj/16day4_orbit4.html (last retrieved 06.01.2020).

3 Roger Revelle, Hans E. Suess, 'Carbon Dioxide Exchange between Atmosphere and Ocean and the Question of an Increase of Atmospheric CO2 during the Past Decades', in: *Tellus*. Informa UK Limited, 9 (1): pp. 18–27.

4 Carbon Dioxide Information Analysis Center: 'Since 1751 just over 400 billion metric tonnes of carbon have been released to the atmosphere from the consumption of fossil fuels and cement production. Half of these fossil-fuel CO2 emissions have occurred since the late 1980s.' https://cdiac.essdive.lbl.gov/trends/emis/tre_glob_2014.html (last retrieved 06.01.2020).

Chapter Three: Nature and Life

1 Joseph A. Tainter, *The Collapse of Complex Societies*, Cambridge University Press, 1988, p. 50.

2 Cf., for example, 'Kein Mensch will Tiere am ersten Tag töten', *Tagesspiegel*, 31.03.2015, https://www.tagesspiegel.de/wirtschaft/gegenkuekenschreddern-kein-mensch-will-tiere-am-ersten-tag-toeten/11578688.html, or 'Das Gemetzel geht weiter', *Süddeutsche Zeitung*, 29.02.2018, https://www.sueddeutsche.de/wirtschaft /kuekenschreddern-das-gemetzel-geht-weiter1.3924618 (last retrieved 06.01.2020).

3 See 'Burning Deadstock? Sadly, "Waste is nothing new in fashion"', Fashion United, 19.10.2017, https://fashionunited.uk/news/fashion/burning-apparel-deadstock-sadly waste-is-nothingnew-in-fashion/2017101926370 (last retrieved 06.01.2020).

4 *Our Common Future, The World Commission on Environment and Development*, Commission for the Future, Oxford University Press, 1990. https://sustainabledevelopment.un.org/content/documents/5987our-common-future.pdf (last retrieved 18.05.2022).

5 Robert Solow, 'The Economics of Resources or the Resources of Economics', in: *The American Economic Review*, 1974, Vol. 64, No. 2, pp. 1–14, this quote p. 11.

6 See Federal Agency for Nature Conservation (Bundesamt für Naturschutz) (Ed.), 'Caring for Pollinators. Safeguarding agro-biodiversity and wild plant diversity' (last retrieved 06.01.2020). https://www.bfn.de/sites/default/files/2021-07/skript250.pdf.

7 Robert Costanza, Rudolf de Groot, Paul Sutton, Sander van der Ploeg, Sharolyn J. Anderson, Ida Kubiszewski, Stephen Farber, R. Kerry Turner, 'Changes in the global value of ecosystem services', in: *Global Environmental Change*, Vol. 26, 2014, pp. 152–158.

Chapter Four: Humans and Behaviour

1 John Robert McNeill, quoted in Jeremy Lent, *The Patterning Instinct*, Prometheus Books, 2017, p. 398.

2 See James Gamble, 'The Most Important Problem in the World', *Medium*, 13.03.2019, https://medium.com/@jgg4553542/the-most-important-problem-in-the-world-ad22ade0ccfe (last retrieved 06.01.2020).

3 This quote appears in E.F. Schumacher's book *Good Work* (1979). It can also be found on the homepage of the Schumacher Institute, https://www.schumacherinstitute.org.uk/about-us/ (last retrieved 16.01.2020).

Chapter Five: Growth and Development

1 Joseph Stiglitz, 'It's time to retire metrics like GDP. They don't measure everything that matters', *The Guardian*, 24 November 2019.

2 Henrik Nordborg, 'Ein Gespenst geht um auf der Welt—das Gespenst der Fakten', https://nordborg.ch/wp-content/uploads/2019/05/Das-Gespenst-der-Fakten.pdf (last retrieved 06.01.2020).

3 Cf. Umweltbundesamt [German Federal Environment Agency] (Ed.), 'Stromverbrauch' ['Electricity Consumption'], 03.01.2020, https://www.umweltbundesamt.de/daten/energie/stromverbrauch, and Umweltbundesamt (Ed.), 'Energieverbrauch nach Energieträgern, Sektoren und Anwendungen' [Energy Consumption by Energy Source, Sector and Use], 03.01.2020, https://www.umweltbundesamt.de/daten/energie/energieverbrauch-nach-energietraegern-sektoren (last retrieved 06.01.2020).

4 Cf. Ernst Ulrich von Weizsäcker, Andus Wijkman et al. *Come On!: capitalism, short-termism, population and the destruction of the planet*, Springer, 2018.

5 *The Wealth of Nations*, MetaLibri, 2007, p. 13.

6 Cf. Jason Hickel, 'Bill Gates says poverty is decreasing. He couldn't be more wrong', *The Guardian*, 29.01.19, https://www.theguardian.com/commentisfree/2019/jan/29/bill-gates-davos-global-poverty-infographic-neoliberal (last retrieved 06.01.2020).

7 David Woodward, '*Incrementum ad Absurdum*: Global Growth, Inequality and Poverty Eradication in a Carbon-Constrained World', World Social and Economic Review 2015, No. 4.

8 Jan Göbel, Peter Krause, 'Einkommensentwicklun g—Verteilung, Angleichung, Armut und Dynamik' ('Developments in Income—Distribution, Alignment, Poverty and Change'), *Destatis Datenreport*, 2018, pp. 239–253, https://www.destatis.de/DE/Service/Statistik-Campus/Datenreport/Downloads/datenreport-2018-kap-6.pdf?__blob=publicationFile (last retrieved 06.01.2020).

9 World Inequality Lab, *World Inequality Report* 2018, English version, p. 11, https://wir2018.wid.world/files/download/wir2018-summary-english.pdf (last retrieved 06.12.2019).

10 Gabor Steingart, 'Konzerne manipulieren nach Belieben die Aktien—und der Staat schaut einfach zu' ('Corporations manipulate shares at will while the state simply looks on'), *Focus Online, Finanzen100*, 08.11.2019, https://www.finanzen100.de/finanznachrichten/boerse/konzerne-manipulieren-nach-belieben-die-aktienkurse-und-derstaat-schaut-einfach-zu_H1907961083_11325544/ (last retrieved 16.01.2020).

11 *Tagesschau*, 'Milliarden für die Aktionäre: Geldmaschine JPMorgan' ('Billions for the Shareholders: the JPMorgan money machine'), boerse.ard.de, 16.07.2019, https://www.tagesschau.de/wirtschaft/boerse/jpmorgan-gewinne-101.html (last retrieved 16.01.2020).

12 Linsey McGoey, 'Capitalism's Case for Abolishing Billionaires', *Evonomics*, 27.12.2019, https://evonomics.com/capitalism-case-for-abolishing-billionaires/ (last retrieved 16.01.2020).

13 'Neue Wert-Schöpferin' ('New Value(s) Creator'), *Manager Magazin*, 08/2018, https://heft.manager-magazin.de/MM/2018/8/158462586/index.html (last retrieved 06.01.2020).

Chapter Six: Technology and Progress

1 Jeremy Lent, *The Patterning Instinct*, Prometheus Books, 2017, p. 378.

2 Jevons, W. Stanley, *The Coal Question*, London, 1865, p. 102. Online edition: https://archive.org/details/in.ernet.dli.2015.224624/page/n123/mode/2up?view=theater (last retrieved 08.06.2022).

3 Cf. Uwe Schneidewind, *Die Große Transformation* (*The Great Transformation*), Frankfurt am Main, 2018, p. 58.

4 IVL Swedish Environmental Research Institute, *Lithium-Ion Vehicle Battery Production. Status 2019 on Energy Use, CO2 Emissions, Use of Metals, Products Environmental Footprint, and Recycling*, https://www.ivl.se/download/18.14d7b12e16e3c5c36271070/1574923989017/C444.pdf (last retrieved 14.07.2020).

5 Tim Jackson, Peter A. Victor, 'Unraveling the claims for (and against) green growth', *Science Magazine*, 22.11.2019, https://www.science.org/doi/abs/10.1126/science.aay0749?doi=10.1126/science.aay0749.

6 Holger Holzer, 'Tesla Cybertruck in Europa möglicherweise nicht zulassungsfähig' ('Tesla Cybertruck Possibly Ineligible for Approval in Europe'), *Handelsblatt*, 16.12.2019, https://www.handelsblatt.com/mobilitaet/motor/elektro-pickup-tesla-cybertruck-in-europa-moeglicherweise-nicht-zulassungsfaehig/25338516.html.

7 Philipp Staab, *Falsche Versprechen* (*False Promises*), Hamburg, 2016, pp. 75–76.

8 Georg Franck, *Ökonomie der Aufmerksamkeit* (*The Economy of Attention*), Munich, 1998; https://www.heise.de/tp/features/The-Economy-of-Attention-3444929.html.

9 Cf. Douglas Rushkoff , 'We shouldn't blame Silicon Valley for technology's problems—we should blame capitalism', *Quartz*, 24.01.2019, https://qz.com/1529476/we-shouldnt-blame-silicon-valley-for-technologysproblems-we-should-blame-capitalism/ and The Associated Press, 'Ex-Google exec Harris on how tech downgrades humans', *sentinel*, 11.8.2019, https://sentinelcolorado.com/sentinel-magazine/qa-ex-google-exec-harris-on-how-tech-downgrades-humans/ (last retrieved 06.01.2020).

Chapter Seven: Consumption

1 Robert Quillen, 'Paragraphs', *The Detroit Free Press*, 4 June 1928.

2 Stephan Lessenich, *Living Well at Others' Expense: the hidden costs of Western prosperity*, Cambridge, 2019.

3 Cf. https://www.aeb.com/media/docs/press-de/2019-10-02-pressemeldung-aeb-esd-abfallexporte.pdf (last retrieved 16.01.2020), and https://www.handelsblatt.com/unternehmen/handel-konsumgueter/abfall-deutschland-exportiert-mehr-muell-als-maschinen/25078510.html (last retrieved 16.01.2020).

4 Cf. Heinrich Böll Foundation, Institute for Advanced Sustainability Studies (Ed.), *Soil Atlas; Facts and Figures about Earth, Land and Fields*, https://www.boell.de/sites/default/files/soilatlas2015_ii.pdf (last retrieved 06.01.2020).

5 See also Hartmut Rosa, *The Uncontrollability of the World*, Polity Press, 2020.

6 Barry Schwartz, *The Paradox of Choice*, HarperCollins, 2004.

7 Tim Kasser, *The High Price of Materialism*, Bradford Books/The MIT Press, 2002.

8 Derek Curtis Bok, *The Politics of Happiness: what government can learn from the new research on well- being*, Princeton, 2010, p. 15.

9 Armin Falk, 'Ich und das Klima' ('The Climate and Me'), *Die Zeit*, 21.11.2019, https://www.zeit.de/2019/48/klimaschutz-klimawandel-egoismus-verhaltensoekonomie.

Chapter Eight: The Market, the State, and the Common Good

1 'Complexity Economics Shows Us Why Laissez Faire Economics Always Fails', *Evonomics*, 21 February 2016.

2 Cf. Heinrich Böll Foundation (Ed.), 'Fünf Konzerne beherrschen den Weltmarkt' (Five Corporations Dominate the Global Market), https://www.boell.de/sites/default/files/agrifoodatlas2017_facts-and-figures-about-the-corporations-that-control-what-we-eat.pdf.

3 Cf. the GDP figures on the World Bank's web pages https://data.worldbank.org/indicator/NY.GDP.MKTP.CD?view=map.

4 Mariana Mazzucato, *The Entrepreneurial State: debunking public vs. private sector myths*, Penguin Books, 2018.

5 Cf. 'The Silicon Six', *Fairtaxmark*, December 2019, https://fairtaxmark.net/wp-content/uploads/2019/12/Silicon-Six-Report-5-12-19.pdf (last retrieved 06.01.2020).

6 Cf. 'Amazon in its Prime', Institute on Taxation and Economic Policy (ITEP), 13.02.2019, https://itep.org/amazon-in-its-prime-doubles-profits-pays-0-in-federal-income-taxes/ (last retrieved 06.01.2020).

7 Karen Vaughn, *The Invisible Hand*, Palgrave Macmillan, 1989, p. 171.

8 John Maynard Keynes, 'The End of *Laissez-Faire*', the Hogarth Press, 1926, p. 19.

9 Cf. University of Bamberg (Ed.) 'Präventives Retourenmanagement und Rücksendegebuehren—Neue Studienergebnisse' ('Preventive Returns Management and Returns Fees—New Research Results'), retourenforschung.de, press release from 11.02.2019, http://www.retourenforschung.de/info-praeventives-retourenmanagement-und-ruecksendegebuehren---neue-studienergebnisse.html (last retrieved 06.01.2020).

10 A transcript of the 'fireside chat' is available online at: https://teachingamericanhistory.org/document/fireside-chat-on-the-new-deal/ (last retrieved 06.01.2020).

11 See article 14, paragraph 2.

12 Thomas Beschorner, *In schwindelerregender Gesellschaft* (*In Dizzying Company*), Murmann Verlag, 2019.

13 For the market share of organic food, see https://de.statista.com/statistik/daten/studie/360581/umfrage/marktanteil-von-biolebensmitteln-in-deutschland; https://www.statista.com/statistics/632803/share-of-organic-product-groups-germany/. For the market share of organically farmed meat, see https://www.fleischwirtschaft.de/wirtschaft/nachrichten/Bio-Markt-Der-Umsatz-waechst-38580?crefresh=1 (last retrieved 06.01.2020).

14 The *Lancet Planetary Health* (Ed.), 'More than a Diet', February 2019, Vol. 3, Issue 2, https://www.thelancet.com/journals/lanplh/article/PIIS2542-5196%2819%2930023-3/fulltext (last retrieved 06.01.2020).

15 For spending on food, see https://de.statista.com/statistik/daten/studie/75719/umfrage/ausgaben-fuer-nahrungsmittel-indeutschland-seit-1900/; https://www.statista.com/statistics/515276/food-consumer-spending-in-germany/; for developments in housing costs, see https://makronom.de/wie-die-veraenderung-der-wohnausgaben-die-ungleichheit-erhoeht-hat-28291 (last retrieved 06.01.2020).

Chapter Nine: Fairness

1 Anand Giridharadas in a speech to the Aspen Institute's Action Forum, 29 July 2015, 'The Thriving World, the Wilting World, and You', *Medium*

2 Stefan Gössling, 'Celebrities, air travel, and social norms', *ScienceDirect*, No. 79, November 2019, https://www.sciencedirect.com/science/article/abs/pii/S016073831930132X (last retrieved 06.01.2020).

3 The so-called carbon dioxide clock gives constantly updated figures on our remaining carbon budget: https://www.mcc-berlin.net/en/research/co2-budget.html (last retrieved 06.01.2020).

4 The Forbes list can be found here: https://www.forbes.com/billionaires/#36ccf2b9251c (last retrieved 06.01.2020).

5 Cf. Dan Ariely, 'Americans Want to Live in a Much More Equal Country', *The Atlantic*, 02.08.2018, https://www.theatlantic.com/business/archive/2012/08/americanswant-to-live-in-a-much-more-equal-country-they-just-dont-realizeit/260639/; and http://danariely.com/2010/09/30/ (last retrieved 06.01.2020).

6 These figures are taken from the work of the economist and inequality researcher Gabriel Zucman, summarised in Pedro da Costa, 'Wealth Inequality Is Way Worse Than You Think, And Tax Havens Play A Big Role', *Forbes*, 12.02.2019, https://www.forbes.com/sites/pedrodacosta/2019/02/12/wealth-inequality-is-way-worse-than-you-think-and-tax-havens-play-a-big-role/#1672b3ceeac8 (last retrieved 06.01.2020).

7 A summary of the report can be found at https://wir2018.wid.world/files/download/wir2018-summary-english.pdf (last retrieved 06.01.2020).

8 Ibid.

9 Cf. Information at *Forbes* https://www.forbes.com/sites/

pedrodacosta/2019/02/12/wealth-inequality-is-way-worse-than-you-think-and-tax-havens-play-a-big-role/#1672b3ceeac8 (last retrieved 06.01.2020).

10 Cf. Mark Curtis, *Gated Development: is the Gates Foundation always a force for good?*, Global Justice Now (Ed.), June 2016, https://www.globaljustice.org.uk/sites/default/files/files/resources/gjn_gates_report_june_2016_web_final_version_2.pdf (last retrieved 06.01.2020).

11 Cf. Giridharadas' talk at the Aspen Institute's Action Forum, 29 July 2015: Anand Giridharadas, 'The Thriving World, the Wilting World, and You', *Medium*, 01.08.2015, https://medium.com/@AnandWrites/the-thriving-world-the-wilting-world-and-you-209ffc24ab90 (last retrieved 06.01.2020).

12 Jeff Cox, 'CEOs see pay grow 1,000% in the last 40 years, now make 278 times the average worker', *CBNC*, 16.08.2019, https://www.cnbc.com/2019/08/16/ceos-see-pay-grow-1000percent-and-now-make-278-times-the-average-worker.html (last retrieved 06.01.2020).

13 https://wriorg.s3.amazonaws.com/s3fs-public/uploads/historical_emissions.png

14 Cf. Helmholtz Centre for Environmental Research (Ed.) 'The Forests of the Amazon are an important carbon sink', 08.11.2019, https://www.ufz.de/index.php?en=36336&webc_pm=48%2F2019 (last retrieved 06.01.2020).

15 Claudia Krapp, 'Waldbrände mit ›ungewöhnlichen‹ Folgen' ('Forest Fires with 'Unusual' Consequences'), *Forschung und Lehre*, 15.10.2019 https://www.forschung-und-lehre.de/forschung/waldbraende-mit-ungewoehnlichen-folgen-2213/ (last retrieved 06.01.2020).

16 Cf. M. Shahbandeh, 'Trade Value of Leading Beef Exporters Worldwide in 2020', *Statista* 15.12.2021 https://www.statista.com/statistics/917207/

top-exporters-of-beef-global/ (last retrieved 27.06.2022).

17 Cf. 'Infografiken Sojawelten: Die Zahlen' ('Infographics Soya-Worlds: The Figures'), *transgen*, last update 20.03.2019, https://www.transgen.de/lebensmittel/2626.soja-welt-zahlen.html (last retrieved 06.01.2020).

18 Ha-Joon Chang, *Kicking away the Ladder: development strategy in historical perspective*, Anthem Press, 2002, p. 129.

19 Oliver Richters, Andreas Siemoneit, *Marktwirtschaft reparieren*, Munich 2019, p. 158.

20 Daniel Markovitz, 'How Life Became an Endless, Terrible Competition', *The Atlantic*, September 2019, https://www.theatlantic.com/magazine/archive/2019/09/meritocracys-miserable-winners/594760/ (last retrieved 06.01.2020).

21 Peter Barnes et al., 'Creating an Earth Atmospheric Trust', *Science*, No. 319, 08.02.2008, pp. 724–726, https://www.uvm.edu/~msayre/EAT.pdf.

22 Michael Sauga, 'Forscher halten Systemwechsel für nötig' ('Researchers believe a systemic change is necessary'), *Spiegel*, 12.07.2019, https://www.spiegel.de/wirtschaft/soziales/klimasteuer-der-co2-preis-soll-nicht-die-staats-kasse-fuellen-a-1276939.html ((last retrieved 06.01.2020).

23 Agora Energiewende und Agora Verkehrswende, 'Die Kosten von unterlassenem Klimaschutz für den Bundeshaushalt 2018' ('The cost for the 2018 German national budget of not protecting the climate'), https://www.stiftung-mercator.de/media/downloads/3_Publikationen/2018/Oktober/142_Nicht-ETS-Papier_WEB. pdf (last retrieved 06.01.2020).

Chapter Ten: Thought and Action

1 Maria Popova, 'How We Spend Our Days Is How We Spend Our Lives: Annie Dillard on choosing presence over productivity', https://www.themarginalian.org/2013/06/07/annie-dillard-the-writing-life-1/